Climate Change

Debra D. Joseph · Maud Mthembu ·
Camille Huggins
Editors

Climate Change

Intervention Strategies and Sustainable Development in the Global South

palgrave
macmillan

Editors
Debra D. Joseph
The University of the West Indies
Cave Hill, Barbados

Maud Mthembu
University of KwaZulu-Natal
Durban, South Africa

Camille Huggins
The University of the West Indies
St. Augustine, Trinidad and Tobago

ISBN 978-3-031-83036-5 ISBN 978-3-031-83037-2 (eBook)
https://doi.org/10.1007/978-3-031-83037-2

This Palgrave Macmillan imprint is published by the registered company Springer Nature Switzerland AG
The registered company address is: Gewerbestrasse 11, 6330 Cham, Switzerland

This book is dedicated to persons involved in climate change and all that it entails, especially those in the Global South. We appreciate efforts to bring awareness to this topic and we support and encourage solutions to all challenges faced. This book was written to encourage those concerned in making the world a better place and to let them know that they should never give up on hope for a better future despite the climate change crisis. We can share and learn from each other.

Foreword

Climate change is altering natural and human systems across the globe, but the most significant impacts are being experienced in the nations that contribute least to this planetary crisis. In 2021, Camille Huggins and Debra Joseph collaborated on a book about *Gender and Domestic Violence in the Caribbean,* which revealed strong linkages between floods and other climate-related disasters and family violence in the Caribbean Islands. In 2023, these award-winning professionals were joined by Maud Mthembu as they set about to record the voices of social scientists from the parts of the world most vulnerable to climate change and least represented in the scientific literature. This current volume represents those voices and includes case studies and regional analyses of climate change impacts and adaptation strategies written by social workers in the Global South. It reveals the diversity of observed impacts and the challenges ahead based on their observations, experiences, and empirical studies in their respective countries. The chapters highlight the importance of climate change informed social work practitioners who are equipped to work in disaster-prone areas. The chapters collectively underscore the disproportionate effects of climate change on poor communities in nations with developing economies and, particularly, the burden of coping on women.

Several chapters in this book focus on the unique challenges facing Small Island Developing States and their struggle to achieve sustainable development goals. The case studies from Africa highlight the importance of interventions for those who derive their subsistence directly from natural ecosystems, those with pastoral livelihoods, the elderly, and other at-risk populations when social networks and national policies challenge their ability to survive and prosper in territories of their choice. This body of work draws attention to the important role of social work professionals in understanding and anticipating the impacts of climate change and in the development of solutions that do not amplify impacts on marginalised people groups. It provides context for future social science research and communities of practice in the Global South. It can help inform the design of training programmes for psychologists, mental health counsellors, and other social work practitioners who are often considered the most trusted in providing guidance in communities after a major disaster.

This book represents a unique and valuable departure from the biophysical assessment reports that have underpinned global decisions about climate change mitigation and adaptation. It forms a sort of social science bridge between local knowledge and western science that has dominated the climate change literature for over 30 years. It offers a glimpse into the challenges facing social welfare systems under a changing climate in Africa, Latin America, and the Caribbean. The authors assert that the principles of social justice and collective responsibility are central to social work and that social workers have a responsibility to protect vulnerable populations from the negative impacts of climate change.

The work of the editors and chapter authors illustrates a lesser-known component of climate change story, perhaps inspiring other social science scholars and practitioners from the Global South to document their experiences and describe the challenges they face in supporting communities at risk. Their work is important in building awareness of the human dimensions of climate change in vast southern regions of our changing planet.

Virginia Burkett
United States Geological Survey
Reston, USA

Virginia Burkett has published roughly 100 journal articles, book chapters, and reports that focus on wetlands, global change, and low-lying coastal systems. She was a Lead Author of the United Nations Intergovernmental Panel on Climate Change (IPCC) Third, Fourth and Fifth Assessment Reports and the IPCC Technical Paper on Water. She was also a Lead Author of the First, Second and Third US National Climate Assessments. She served as Chair of the $2.6 billion U.S. Global Change Research Program during 2016–2019. She is currently serving as Chief Scientist for Climate and Land Use Change at the U.S. Geological Survey.

Acknowledgements

I would like to express extreme gratitude for the opportunity to work on this book entitled '*Climate Change: Intervention Strategies and Sustainable Development in the Global South*'. This project highlights the urgent realities facing the Global South in the context of climate change. Many countries in this region are currently grappling with rising temperatures, severe flooding, vector-borne diseases, and water scarcity. These challenges significantly impact livelihoods particularly in agriculture along with land use, public health, air quality, and patterns of migration. The primary goal of this book is to build upon the foundation established by the previous volume, *The Impact of Climate Change on Vulnerable Populations: Social Responses to a Changing Environment*. The earlier work focused on how climate change affects vulnerable populations, exploring the social responses to these challenges from a Social Work perspective. In this follow-up volume, we aim to delve deeper, showcasing strategies for sustainable development and resilience in the Global South.

A heartfelt thanks is extended to the contributing authors who provided chapters, insights, and research. These authors were scholars from the University of the West Indies located in Trinidad and Barbados, South Africa, and South America. The chapters in this book explore a diverse range of critical topics, including the psychological impact of natural disasters, food insecurity, migration patterns of displaced persons, African social work intervention perspectives, the role of small business enterprises in addressing climate change, and the innovative applications of Artificial Intelligence (AI) in this field. Together, these contributions provide a comprehensive view of the multifaceted challenges and solutions related to climate change in the Global South.

Special gratitude for the co-editors, Drs. Maud Mthembu and Camille Huggins for their invaluable input and support as peer reviewers and proofreaders for the various chapters. They were committed to the process and shared the vision of what this publication could provide for the continued issues encountered with regard to climate change.

We extend thanks to Rachael Ballard, Global Head and Publisher, Geography and Environment at Palgrave Macmillan, who assisted us in the first phase of this publication from book proposal to acceptance of same. Her guidance was invaluable. Also, Mr. Naveen Dass, Production Editor, who was assigned to this book publication to ensure that we had a quality finished product. He kept in contact throughout the process with reminders to keep the work in focus.

Dr. Joseph—I would personally like to thank my sons and close friends for their support, understanding, and encouragement. This gave me the motivation to complete the task. Time constraints were a challenge and I had to carve out the time to assist in bringing this publication to fruition. I would also like to thank Dr. Roshnie Doon who assisted with the journey of this book in its initial stages. I am extremely grateful to Professor Adele Jones for her input and support in the process.

Dr. Mthembu—My heartfelt thanks go to my children, Langelihle Cebekhulu and Wakhile Mhlongo, for their unwavering support, patience, and understanding during the long hours I spent immersed in academic pursuits. Your encouragement has been my greatest source of strength. I am forever grateful to the Mthembu family for laying a strong foundation that has empowered me to strive and grow professionally, personally, and spiritually. To my colleagues at the University of KwaZulu-Natal, Discipline of Social Work, your intellectual contributions and camaraderie have been a constant source of inspiration and motivation.

Dr. Huggins—For my niece and nephews, hoping you are prepared for the future.

Finally, to the pioneering scholars in this field, your groundbreaking work has not only paved the way but also provided the solid framework for this exploration. Your dedication continues to inspire me and countless others.

We appreciate the readers and anyone who has shown interest in the book entitled '*Climate Change: Intervention Strategies and Sustainable Development in the Global South*' and we have hope for the book's impact on and contribution to the field of Social Work.

Contents

Notes on Contributors

Letetia M. Addison holds a Ph.D. in Mathematics and M.Phil. in Statistics from UWI, St. Augustine Campus, and currently serves as a Project Officer at the University Office of Planning. With over a decade of experience teaching Mathematics, Probability, and Statistics at tertiary level, she actively engages in lecturing, mentoring, and diverse research. Dr. Addison has a robust publication record and specialises in interdisciplinary projects, including AI applications and predictive modelling for climate change and education. She led a winning cross-campus team in the 2022 GRD Climate Resilience Data Challenge, focusing on developing an AI prototype to classify natural disaster risk in the OECS region. Her current research interests focus on applying mathematical and statistical models for sustainability.

Timothy Affonso is a Deputy Dean at the Faculty of Law at the University of the West Indies, St. Augustine Campus. Dr. Affonso is also a Tutor at the Hugh Wooding Law School where he teaches constitutional law and law and legal systems. He is an attorney-at-law with over 18 years' experience. He is a former State Prosecutor and extradition attorney with the Ministry of the Attorney General of Trinidad and Tobago. He is a national scholarship winner and a Chevening Scholar. He holds a Bachelor of Laws Degree from UWI, Cave Hill Campus, a Master of Laws degree, with Distinction, from the University College London and a Doctor of Philosophy in Public International Law and International Human Rights Law from the UWI, Cave Hill Campus. He is a Director and Co-Chair of the Caribbean Centre for Human Rights and a founding member of the International Law Association (Caribbean Branch). His research interests include international human rights with a focus on vulnerable groups, post-colonial laws, and public international law.

Gervan J. Arneaud is a Behavioural Health Therapist at Journey Behavioral Health Clinic and the Assistant Secretary for the Trinidad and Tobago Association of Psychologists. He graduated from the University of Westminster with a Master of Science in Health Psychology and is currently a student in the Master of Social Work programme at the University of the West Indies, St Augustine. He has worked in several areas including research and consulting, patient care, community health outreach and HIV/STI prevention, and counselling. His main interests are in sexual and reproductive health counselling and promotion and is an active researcher in the areas of psychology, psychiatry, and public health.

Lauren Bain is a sustainability practitioner originally from Trinidad and Tobago. Her day-to-day work focuses on the interactions between biodiversity and climate change, and our impacts and dependencies on both. She has a background comprising sustainable development, energy economics, and social sciences and is passionate about driving climate resilience and channelling sustainable finance into Small Island Developing States, especially within the Caribbean region. Lauren believes strongly that you cannot separate people from planet, so is on a mission to create sustainability projects that prioritise valuing both.

Curtis Charles is a Solution Architect and Data Scientist who served at Microsoft Services on the digital transformation of corporate, government, education, and military installations to strengthen operations and financial performance. In addition, Dr. Charles has served as the Vice President for Research, Development, and Innovation at RLM Communications (DoD Contractor), where he oversaw design, development, and innovation of new products and services for Big Data Analytics; Artificial Intelligence (AI); Machine Learning; Biometrics; and Cybersecurity. Director of Academic Affairs at the UWI Five Island Campus, Dr. Charles is responsible for the School of Science, Computing and Artificial Intelligence; School of Management Science and Technology; the School of Health and Behavioral Sciences; the School of Humanities and Education; Institutional Effectiveness; Sponsored Research and Programs. Dr. Charles is a graduate of Massachusetts Institute of Technology

(MIT), Howard University, Morgan State University, and Northwestern University.

Edward Clarke is Barbados born and educated in England has worked extensively across the Caribbean and Southern Africa. He is the founding Dean of the School of Social Science and an Adjunct Professor of Counselling Psychology at the University of the Southern Caribbean (USC) in Trinidad and Tobago. He earned his degrees from prestigious institutions, including the universities of London, Surrey, and Oxford Brookes in England. His post-doctoral qualifications include professional certificates in Cognitive Behavioural Therapy (CBT), Trauma Therapy for Children and Adolescents, Grief Counselling (CG-C), Dialectical Behaviour Therapy (DBT), Online and Telephone Counselling (UK), Emotional Intelligence Consulting (USA), and Clinical Counselling Supervision (UK). Having lived and worked in multiple Southern African countries, Dr. Clarke's African experience spans Lesotho (where he was a Senior Lecturer in Sociology at the University of Roma), Swaziland, Namibia, Botswana, Zimbabwe, and Angola. He has also served as a part-time research consultant for international organisations such as WHO, UNICEF, UNFPA, and USAID. Dr. Clarke has also made scholarly contributions, co-authoring a chapter 'Dying Too Young in the Caribbean' in Legacy of the Crossing, edited by LaVeist, T. A (2017). Additionally, he has authored a chapter entitled 'Exploring Domestic Violence in the Seventh-Day Adventist Church' in the book Domestic Violence in the Anglophone Caribbean, edited by Bissessar, AM, (2021); and co-author of the chapter 'An Overview of Caribbean Burial Rituals' in Post-Colonial Burial and Grieving Rituals of the Caribbean, edited by Huggins C., Bissessar, AM, & Hinkson, GM, (2024).

Nkosiyazi Dube (D.Phil.) holds a BSW, MA (Social Development), and a D.Phil. (Wits). Nkosiyazi Dube is a Lecturer in the Department of Social Work in the School of Human and Community Development at the University of the Witwatersrand. He is registered with the South African Council for Social Service Professions (SACSSP). Dr. Dube's teaching and research interests are health and social development, children and HIV and AIDS, poverty and development, children in alternative care, formal and informal social security, social networks, and health promotion.

Shalee N. Forney (she/her/hers) is an ABD from the Joint Ph.D. Program in Social Work at the North Carolina Agricultural and Technical

State University (NCAT) and University of North Carolina Greensboro (UNCG). Her interests include juvenile justice, restorative practices, and immigrant and refugee communities.

Ashley D. Gonzalez (she/her/hers) is an alum of the Joint Ph.D. Program in Social Work at the North Carolina Agricultural and Technical State University (NCAT) and University of North Carolina Greensboro (UNCG). Her interests include delivering services to immigrants and refugees globally, criminal justice, and sociology.

Dr. Glenda Hinkson is an Assistant Professor and former Chair of the Social Work Program at the University of the Southern Caribbean in St. Joseph, Trinidad, where she served for seven years. She specialises in teaching advanced clinical courses within the Social Work and Psychology Departments. With 17 years of experience as a practising couple and family therapist, Dr. Hinkson has focused her clinical work on supporting HIV-positive and affected youth and adults, as well as working within the family court system in Philadelphia, Pennsylvania. Her research interests encompass loss, death, and grief, with a particular emphasis on the dynamics of families, couples, and relationships across the lifespan. Dr. Hinkson's most recent publications include the co-edited volume Postcolonial Burial and Grieving Rituals of the Caribbean (Huggins, Bissessar, Hinkson, 2024) and the article 'Transnational Caregiving and Grief: An Autobiographical Case Study of Loss and Love During the COVID-19 Pandemic', OMEGA Journal of Death and Dying (Hinkson GM, Huggins CL, Doyle M. (2022)). In addition to her academic and clinical work, Dr. Hinkson holds certifications in trauma and sexual health education, and she is also a certified Health Educator and HIV Counseling and Testing Specialist. Dr. Hinkson earned her Ph.D. in Couple and Family Therapy from Drexel University in Pennsylvania. She also holds a Master's degree in Marriage and Family Therapy from Syracuse University and a Master's in Counselling Psychology from Howard University, all in the USA. She has actively participated in numerous workshops, lectures, and conferences in the USA, Barbados, and Trinidad, sharing her expertise with diverse audiences.

Camille Huggins is a Lecturer in the Department of Behavioural Science at the University of the West Indies, St. Augustine campus in Trinidad and Tobago since 2015. Dr. Huggins obtained her Master's degree from Columbia University School of Social Work and a Ph.D. from New York University. Dr. Huggins's research interest is traumatic experiences among minoritised women. Dr. Huggins has authored three books including in 2021 and 2022 titled Gender and Domestic Violence in the Caribbean and Domestic Violence in the Anglophone Caribbean.

Debra D. Joseph Trinbago born, is the coordinator of and tenured lecturer in Social Work at the University of the West Indies, Cave Hill, Barbados. Dr. Joseph is also a clinical social worker who has worked with individuals and families for over 14 years. She is involved with the issues of HIV/AIDS that affect women and children from a clinical and research standpoint. She is also an avid supporter and researcher of social work's involvement in climate change, environmental sustainability and domestic violence and has published in these areas. Dr. Joseph supports the None in Three EU initiative on Domestic Violence at the University of Huddersfield and has analysed the men and youth aspect of the qualitative research segment for Barbados and Grenada. She is presently an honorary consultant at that establishment. She has recently completed research on the 'Women Fisher Folk Barbados-A Gendered Lens' Research Project and 'Female Headed Households: Aftermath of Hurricane Maria'. She is also the recipient of the Jeremy Collymore Award for Research in Humanitarian Response and Disaster Risk Management 2019. This research has culminated in a book chapter for her first published edited book entitled 'Climate Change and Vulnerable Populations'. Dr. Joseph had also won a scholarship to attend the 'Women Deliver' Conference in Vancouver in 2019 because of her research that shed light on gender issues.

Priscalia Khosa is an Associate Professor in the Department of Social Work and Community Development at the University of Johannesburg, South Africa. She published widely on her fields of research interests, which include social work supervision, social work education, and substance abuse. She has supervised postgraduate research on different fields which include green social work among others.

Sara Low is a Trinidadian environmental advocate passionate about climate resilience in small islands. With an academic foundation in environmental biology and professional experience in corporate sustainability,

Sara understands both the science and strategic action needed to address the climate crisis. Serving as the 2024–2025 Curator of Global Shapers Port-of-Spain, a local youth-led social justice organisation, Sara actively works to drive positive change in the Caribbean through community projects and advocacy. Through her global volunteer and work experiences, she recognises the value of collaboration across multiple sectors to address the challenges associated with the climate crisis. Believing deeply in the power of collective action, Sara aims to inspire others to embrace their roles in creating a sustainable and resilient future for Trinidad and Tobago.

Ken Manohar has worked in the IT Service industry for the past twenty (20) years. His experience is varied, from running network cables, to implementing email clusters, to creating hybrid environments using SaaS solutions. He is currently pursuing the M.Sc. Data Science at UWI, St. Augustine.

Poppy Masinga is the Head of Faculty of Social Work and Community Development at South African College of Applied Psychology (SACAP); Current President of the Association of South African Social Work Education Institutions (ASASWEI); Current Chairperson of International Association of Social Work with Groups (IASWG)—Africa/South African Chapter. She is a member of FIRE Network—an international forum of female researchers from different disciplines. A Continuous Professional Development (CPD) and Foreign Qualifications Panel Member of SACSSP. She was former Social Work Practitioner within the Department of Social Development (Gauteng Province) and NGO Sector in various fields. Currently, Dr. Poppy serves on Editorial Boards of Journal of Evidence-Based Social Work and Journal of International Migration and Integration, and Social Work/Maatskaplike Werk. She has special interest in youth development, community development programme design and management and intervention research. She is a social, economic and environmental justice advocate, fighting for the rights of vulnerable groups, including immigrants. She is passionate about green social work, fighting for the protection of plants and animals.

Carmen C. Monico (they/them/elle) is an Associate Professor in the Joint Programs in Social Work at the North Carolina Agricultural and Technical State University (NCAT) and University of North Carolina Greensboro (UNCG) and Scholar of various U.S. Fulbright Programs.

They are engaged in global migration, human trafficking, intercountry adoption, and community engagement research.

Sadia Moore is a citizen of St. Lucia, who currently serves as Chair of the Department of Psychology, Behavioral Sciences, Criminology and Sociology, in the School of Social Sciences at the University of the Southern Caribbean, Trinidad and Tobago. Ms. Moore received her Bachelor of Science Degree in Behavioral Sciences (Sociology emphasis), and a Master of Arts Degree in Educational Psychology from Andrews University, Michigan. Additionally, she is certified Nutrition & Wellness Coach. She has been lecturing at the University of the Southern Caribbean for the past ten (10) years, an experience that she continues to enjoy.

Maud Mthembu is a Senior Lecturer in the Discipline of Social Work within the School of Applied Human Sciences at the University of KwaZulu-Natal (UKZN). As a practicum coordinator for undergraduate studies, she provides strategic leadership to the Discipline social work practicum. Over the past six years, Dr. Mthembu has directed international teaching, learning, and research initiatives aimed at strengthening social work competencies with local and international universities. In addition, she leads an international research project focusing on trauma in HIV cure research with ATIs. Dr. Mthembu's contributions are widely recognised. In 2024, she was nominated for the UKZN Best Teacher's Award, while in 2021 she was honoured as a runner-up for the ASASWEI Best Researcher Award and was named Best Emerging Scholar in 2016. Additionally, she serves on the Board of the Caribbean Journal of Social Work. Her research interests focus on child-centred practices for epidemic response, child protection, and children's mental health.

Tamsin Leigh Nel is a Social Work Masters' graduate from Stellenbosch University. The focus of her research is on green social work as a professional response to climate change from the perspective of social workers. She is also employed as an adoption social worker, working for a non-governmental organisation in Somerset West, Cape Town.

Sthembiso Phoswa is in the final stages of completing his Master's thesis. He is a social work mentor and a practitioner at Ma'at Institute and has over three years of work experience. At Ma'at, he provides individual and group services and training on African-centred psychosocial support, trauma debriefing and African community healing and resilience techniques. His research interests are decoloniality, Afrocentric social work, and *ubuntu* as a philosophy, value system, and practice for building communities and social cohesion.

Kevan Rajaram is a data professional for the past ten (10) years. He has worked at major companies throughout the Caribbean region and most recently at PwC Caribbean as the Senior Manager, Data & Advanced Analytics. He has graduated from the University of the West Indies with a B.Sc. Mathematics (Honours) and a M.Sc. Statistics (Distinction). He is currently pursuing his Ph.D. in Computer Science at UWI, St. Augustine, where his research area is in applications of Artificial Intelligence and Machine Learning.

Brenda Nyeverwai Rumutsa is a Researcher at Parliament of Zimbabwe. She has a doctorate degree in Environmental Science (Geography) from University of Venda (RSA). She also holds a Master's in Disaster Management from NUST (ZW), an Honours in Environmental Science (Land Conservation and Reclamation) from Bindura University of Science Education, as well as a certificate in Occupational Safety, Health, and Environmental Management (OSHEMAC) from National Social Security Authority (NSSA-ZW). From 2014 to 2017, she worked as a Lecturer before pursuing Ph.D. studies and research work in academic and professional projects. Her professional interests focus on research work on disaster and environmental management issues.

Molly Shayamano is a qualified social worker and emerging researcher. She is currently working at the Eastern Cape Department of Education. She holds a BSW from the University of Zimbabwe, a MSW from University of Fort Hare, and a Ph.D. in Social Work from the University of the Witwatersrand. Dr. Shayamano is on a path to become extensively published in her field. She has published in accredited journals and presented at international conferences. She has interest in wide research areas which include green social work, community development, poverty, unemployment issues, and climate change. She is also a member of the Association of South African Social Work Education Institutions' Special Interest Groups on Environmental social work.

Melony Shoye has a B.Sc. (Hons) in Social Work and an M.Sc. in Natural Resource and Environmental Management, both gained from the University of the West Indies, Cave Hill Campus. She is a Barbados Welfare Officer with 21 years of experience and has an interest in issues surrounding the elderly, electronic mapping and disaster management.

She is currently pursuing an M.Phil. in Social Work and has completed same.

Helen B. Tadese (she/her/hers) is a doctoral candidate in the Joint Ph.D. Program at the North Carolina Agricultural and Technical State University (NCAT) and the University of North Carolina Greensboro (UNCG). Her research interests are the mental health of immigrants and refugees and the role of coping mechanisms, resiliency, self-image, and public stigma on these populations.

Sibonsile Zibane is a Senior Lecturer and the Academic Leader in the Discipline of Social Work at the University of KwaZulu-Natal, South Africa. She has 30 years of diverse social work-related experience. Her research interests are Afrocentric social work, community-based work with indigenous communities, gender and sexuality, and trauma-informed care. She is also a project specialist and one of the founding members of the Ma'at Institute, whose vision is pioneering African-centred psychosocial community services and scholarship. The Institute has become a vital role player in disaster management, mitigating the impact of COVID-19, and multidisciplinary social and mental health interventions embedded in the African paradigm. Furthermore, it provides learnership opportunities to unemployed graduates in the fields of social work, psychology, social sciences, law, communication, and media studies.

Nokukhanya Zondi is a social work mentor, a practitioner, and an *IsiZulu* language expert at Ma'at Institute, which is in the School of Applied Human Sciences, University of KwaZulu-Natal, South Africa. At Ma'at, she provides training on African-centred psychosocial interventions. She has more than three years of work experience. MaZondi's research interests are indigenous African knowledge systems, African-centred psychosocial interventions, African culture, and Afrocentric social work. Her highest qualification is a Master of Laws degree in child protection. Her Master's research explored child participation in African indigenous families. She is currently a registered Ph.D. student in the Discipline of Social Work. Her Ph.D. study focuses on rites of passage and rituals for childhood development in indigenous African communities. She is writing and conducting her Ph.D. using an *IsiZulu* indigenous language.

Abbreviations

AARs	After Action Reviews
AI	Artificial Intelligence
AIC	African Indigenous Communities
AOSIS	Alliance of Small Island States
APA	American Psychological Association
API	African-centred Psychosocial Interventions
CCCCC	Caribbean Community Climate Change Centre
CDEMA	Caribbean Disaster Emergency Management Agency
COP	Conference of the Parties
COVID	Coronavirus Disease
CPD	Continuous Personal Development
DRSmC	Disaster Risk Smart Classifier
DSM	Diagnostic and Statistical Manual of Mental Disorders
DSW	Developmental Social Welfare
EM	Ethekwini Municipality
EPA	Environmental Protection Agency
FAO	Food and Agricultural Organisation
GATE	Government Assistance for Tuition Expenses
GDP	Gross Domestic Product
GHG	Green House Gases
GRID	Global Report on Internal Displacement
HIV/AIDS	Human Immunodeficiency Virus/Acquired Immunodeficiency Syndrome
ICT	Information and Communication Technology
IDMC	Internal Displacement Monitoring Centre
IFRC	International Federation of Red Cross and Crescent Societies

IFSW	International Federation of Social Work
IPCC	Intergovernmental Panel on Climate Change
ISDM	Integrated Service Delivery Model
KZN	KwaZulu-Natal
LAC	Latin America and the Caribbean
LGBT	Lesbian, Gay, Bisexual, Transgender and Intersex
LR	Logistic Regression
LWC	Living Waters Community
ML	Machine Learning
MSMEs	Micro, Small, and Medium Enterprises
MYDNS	Ministry of Youth Development and National Service
NAPA	National Adaptation Programme of Action
NASA	The National Aeronautics and Space Administration
NCCAS	National Climate Change Adaptation Strategy
NCCRP	National Climate Change Response Plan
NDC	Nationally Determined Contribution
NGO	Non-Governmental Organizations
NHC	National Hurricane Centre
NOAA	National Oceanic and Atmospheric Administration
OCHA	United Nations Office for the Coordination of Humanitarian Affairs
OECS	Organization of Eastern Caribbean States
PAHO	Pan-American Health Organization
PANCAP	Pan Caribbean Partnership Against HIV and AIDS
PMPRRR	Prevention, Mitigation, Preparedness, Response, Recovery, and Rehabilitation
POC	Proof-of-Concept
PTSD	Post-Traumatic Stress Disorder
PVO	Private Voluntary Organisations
SDGs	Sustainable Development Goals
SIDS	Small Island Developing States
SMEs	Small and Medium Enterprises
SVM	Support Vector Machine
T&T	Trinidad and Tobago
TCDPO	Town and Country Development Planning Office
UN	United Nations
UNAIDS	Joint United Nations Programme on HIV/AIDS
UNCC	United Nations Climate Change
UNCTAD	United Nations Conference on Trade and Development
UNDESA	United Nations Department of Economic and Social Affairs
UNDP	United Nations Development Programme
UNDRR	United Nations Office for Disaster Risk Reduction

UNECLAC	United Nations Economic Commission for Latin America and the Caribbean
UNESCO	United Nations Educational, Scientific and Cultural Organisation
UNFCC	United Nations Framework Convention on Climate Change
UNHCR	United Nations High Commissioner for Refugees
UNICEF	United Nations International Children's Emergency Fund
USD	United States Dollars
UWI	University of the West Indies
WHO	World Health Organisation
WMO	World Meteorological Organisation

List of Figures

List of Tables

CHAPTER 1

Introduction

Debra D. Joseph, Camille Huggins, and Maud Mthembu

According to Dados and Connell (2012), the phrase 'Global South' refers broadly to the regions of Latin America, Africa, Asia, and Oceania. It denotes regions outside of Europe and North America that are mostly (not all) low-income and which are often politically or culturally marginalised. The Global South, home to most of the world's population and some of its most biodiverse regions, stands at the frontline of the climate crisis. These nations, with complex socio-economic landscapes and historical underpinnings of colonial exploitation, now face escalating environmental pressures. Rising temperatures, erratic weather patterns, and intensified natural disasters pose unprecedented challenges, impacting

D. D. Joseph (✉)
The University of the West Indies, Cave Hill, Barbados
e-mail: debra.joseph@uwi.edu

C. Huggins
The University of the West Indies, St. Augustine, Trinidad and Tobago
e-mail: camille.huggins@uwi.edu

M. Mthembu
University of KwaZulu-Natal, Durban, South Africa
e-mail: mthembum4@ukzn.ac.za

D. D. Joseph et al. (eds.), *Climate Change*,
https://doi.org/10.1007/978-3-031-83037-2_1

livelihoods, destabilising food systems, and forcing communities into increasingly precarious situations.

Strazzante et al. (2021) purports that the Global Climate Risk Index 2021 shows that the poorest countries in the world, even though they may be viewed as having the lowest industrial pollution levels, seem more susceptible to the negative impact of climate change. The inequity experienced by countries in the Global South situates them in a position of disadvantage, exposing them to the vulnerabilities of climate change. Consequently, this exacerbates the already existing global inequalities.

Ogwu et al. (2024) state that many countries in the Global South face significant and growing challenges influencing the attainment of food security. They reiterate that food insecurity in the Global South has an extricable feedback loop linked to poverty, inequality, environmental degradation, biodiversity loss, population growth, and urbanisation, apart from other situations. They add that conflicts, natural disasters, reduction in dietary diversity, and knowledge worsen malnutrition. Diseases, inadequate storage, conflicts, climate change, and poverty-driven migration can impact food insecurity. Consequently, these impacts can increase crime rates, health issues, and political instability.

Donatti et al. (2024) analyse the occurrence of climate-related disasters over 20 years and connected their impact on populations worldwide depending on their levels of human development. They found that Central America, the Caribbean, Eastern Africa, and Southern and Eastern Asia had the highest levels of climate-related disasters. The percentage of persons affected in countries with a high developmental index vastly differed from those with a low human developmental index; the latter with the lower index was more affected. They state that this low human developmental index (HDI) is expected as countries with a high developmental index would likely have more resources to plan for climate-related disasters and warn and educate at-risk populations.

In the Global South, climate change reverberates inequalities by impacting the poorest of communities, for example, Black, Indigenous, and People of Colour, as well as women and children the most. These vulnerable groups can also include smallholder farmers. In many developing countries, farmers produce up to 75% of the food supply. Adverse weather conditions, floods, and droughts affect smallholder farmers' activities (Strazzante et al., 2021).

The following areas highlight some specific adverse climate events:

Eastern Horn of Africa—The UNDRR (2024) report states that there are increasing cycles of flooding and drought in the Eastern Horn of Africa. Ethiopia, Kenya, and Somalia have experienced at least four consecutive failed rainfall seasons. The region had extreme rains in 2019–2020, followed by widespread floods in 2023–2024. The Eastern Horn has experienced eight failed rainy seasons, five average seasons, and three wet seasons due to oceanic influences such as the El Nino. Climate change is increasing these events, leading to devastating rainfall and record temperatures that impact specific sectors such as agriculture, water resources, and health. The report says that environmental degradation reduces the region's capacity to withstand extreme heat, exacerbated by deforestation and land degradation.

As a result of these climate events, shocks are triggered by underlying social factors such as migration, urbanisation patterns, and indebtedness. Over the past three years, from 2020 to 2022, more than 30 million people from Ethiopia, Kenya, and Somalia have faced drought-related food insecurity. Food insecurity included reduced crop and livestock production. In 2023–2024, the worst flood-affected populations were found in the densely populated urban and riverine areas of Somalia, Kenya, and parts of Ethiopia. Consequently, this led to significant food shortages regionally and globally, resulting in high food prices and hyper-inflation. Urbanisation is increasing; for example, in Kenya's Garissa County, 77% of settlements reported more than 205,000 people arriving from other areas because of the drought in their respective countries. Drought could burden Kenya as it copes with its adverse weather events (UNDRR, 2024).

Jamaica (Caribbean)—In 2021, heavy rainfall in Montego Bay flooded roads and damaged vehicles and buildings, affecting approximately 933 people. Montego Bay had experienced flooding as early as 2017, when floods devastated the town centre, flooding businesses, and sweeping cars away (UNDRR, 2024).

Central America—Droughts have been a frequent hazard, impacting ecosystems and life. Since 1981, regional droughts have occurred at least every seven years, lasting about a year and affecting about four countries each time. Climate change in this part of the world is predicted to make dry areas drier and wet areas wetter. There are growing deforestation rates, especially in Honduras and Guatemala (UNDRR, 2024).

Latin America and the Caribbean—Latin America and the Caribbean is the world's second most natural disaster-prone region. Many of these disasters and extreme events are expected to increase in magnitude and frequency due to climate change. Countries are often exposed to natural hazards such as landslides, fires, hurricanes, river and coastal flooding, heat waves, and droughts. During the period 2000–2019, there were 548 floods (the most common recorded disaster), 330 storms, 74 droughts (a disaster that affects the most persons in the region), and 50 extreme weather events. The United Nations Office for Disaster Risk Reduction recorded 175 disasters in the Latin America and Caribbean region for 2020–2022 (UNDRR, 2021).

This book explores the intersections of climate change, intervention strategies, and sustainable development within the Global South. Each chapter tackles a critical dimension of this landscape, examining the urgency of current challenges and the potential pathways towards resilience and equity. The text examines the following key topics:

Disasters—As climate-induced disasters become more frequent and severe, understanding disaster preparedness, mitigation, adaptation, and recovery becomes paramount. This chapter unpacks how communities and governments in the Global South can build adaptive capacities to withstand and recover from disasters while prioritising the protection of vulnerable populations.

The chapters and authors in Part 1 of the book are as follows: *The Impact of Global Boiling on Caribbean Islands: Practicality of Achieving Sustainable Developmental Goals* by Timothy Affonso; *The Psychological Impact of Climate Change: A Recent Overview* by Gervan J. Arneaud; *The Climate-Resilient Elderly of Barbados* by Melony Shoye; *Climate Change-Induced Disasters in Manicaland Province in Zimbabwe: The Need for Continuous Professional Development* by Molly Shayamano, Poppy Masinga, and Nkosiyazi Dube and *Centering the response to the KwaZulu-Natal floods in the African-centred Psychosocial Interventions* by Sibonsile Zibane, Sthembiso Phoswa, and Nokukhanya Zondi.

Food Security—Climate variability intensifies food insecurity, particularly in areas heavily reliant on agriculture. Here, we examine sustainable agricultural practices, technological innovations, and policy frameworks that aim to secure access to food without compromising environmental health.

The chapters and authors in Part II of the book are as follows: *Food Security Through Entomophagy: The Benefits and Challenges of Harvesting Encosternum delegorguei in Climate-Change-Affected Rural Nerumedzo community, Bikita, Zimbabwe* by Brenda Nyeverwai Rumutsa and *Climate Change and Its Effect on the Mental Health of Farmers in Trinidad & Tobago* by Sadia Moore, Glenda Hinkson, and Edward Clarke.

Migration—Climate change is reshaping migration patterns, leading to internal displacement and transnational movement as populations are forced to leave increasingly uninhabitable areas. This chapter explores migration as both a symptom and a coping strategy, investigating policies that could address the root causes and protect the rights of displaced individuals.

The chapters and authors in Part III of the book are as follows: *The Impacts of Climate Change Among Displaced Migrants in Small Island Developing States: The Case of Trinidad and Tobago* by Camille Huggins and *Mitigation and Adaptation Innovations Addressing Climate Change and Forced Migration: Case Study Research in Five World Regions* by Carmen C. Monico, Shalee N. Forney, Helen B. Tadese, and Ashley D. Gonzalez.

Vulnerable Populations, Organisations, and Artificial Intelligence—Climate change and its impacts continue to spotlight vulnerable populations that are most affected. Groups from organisations both private and public and Non-Governmental Organisations (NGOs) are also offering their voices to climate change with cases and possible solutions. Artificial intelligence (AI)'s potential to address climate challenges is immense, yet it is largely underexplored in the context of the Global South. This chapter delves into AI's applications in environmental monitoring and resource optimisation, weighing the benefits against ethical considerations and infrastructural constraints.

The chapters and authors in Part IV of the book are as follows: *Why Climate Change Matters for Vulnerable Populations in South Africa: A Social Work Perspective* by Tamsin Leigh Nel and Priscalia Khosa; *The Power of Small Scale Interventions in Progressing Climate Adaptation and Resilience in the Caribbean: A Case Study of Trinidad and Tobago* by Lauren Bain and Sara Low and *Climate Models Using AI: A Case for the OECS* by Letetia M. Addison, Kevan Rajaram, Ken Manohar, and Curtis Charles.

The Disaster Avoided Model (A work in progress) (UNDRR, 2024) looks at what can be done to build resilience and avoid future disasters and summarises six points that can help, namely:

- The right mindset—Understanding of disasters.
- The suitable investment—Investments are made wisely and at the right time.
- Good Governance—There must be political and economic commitments to manage investments effectively and to deliver tangible social, environmental, and economic benefits.
- Good Data—To inform decisions, reliable data must be gathered through various means, including satellite observations. AI and on-the-ground monitoring.
- Meaningful Targets—Targets can be related to the Sendai Framework, the Sustainable Development Goals (SGDs), climate change efforts, the New Urban Agenda, and the Humanitarian Agenda.
- Meaningful Inclusion—Include all stakeholders, which will help build a society that is resilient to natural forces.

Throughout this book, we advocate for approaches rooted in justice and sustainability, emphasising the importance of including local voices in policy creation and implementation. By focusing on intervention strategies suited to specific contexts within the Global South, this work aims to present actionable insights and foster a deeper understanding of the intricate dynamics at play.

References

Dados, N., & Connell, R. (2012). The global south. *Contexts, 11*(1), 12–13.

Donatti, C. I., Nicholas, K., Fedele, G., Delforge, D., Speybroeck, N., Moraga, P., Zvoleff, A., et al. (2024). Global hotspots of climate-related disasters. *International Journal of Disaster Risk Reduction, 108*, 104488.

Ogwu, M. C., Izah, S. C., Ntuli, N. R., & Odubo, T. C. (2024). Food security complexities in the global south. In *Food safety and quality in the global south* (pp. 3–33). Springer Nature Singapore.

Strazzante, E., Rycken, S. & Winkler, V. (2021). Climate north and global south: How climate change uncovers global inequalities. *Generation Change Europe*. https://gceurope.org/global-north-and-global-south-how-climate-change-uncovers-global-inequalities/

United Nations Office for Disaster Risk Reduction (UNDRR). (2021). *Regional Assessment Report of Disaster Risk in Latin America and the Caribbean: Challenges in Disaster Risk Reduction and Progress Towards the Sendai Framework for Disaster Risk Reduction* (2015–2030).

United Nations Office for Disaster Risk Reduction. (2024). *Forensic insights for future resilience*. Learning from Past Disasters.

PART I

Disasters

CHAPTER 2

The Impact of Global Boiling on Caribbean Islands: Practicality of Achieving Sustainable Development Goals

Timothy Affonso

Introduction

The relationship between climate change and small island developing states (SIDS) is a tangible and sensitive one. The islands of the Commonwealth Caribbean are all SIDS and their vulnerability to the effects of climate change cuts through social and economic distinctions. As climate change worsens and the United Nations has changed the nomenclature from "global warming" to "global boiling", one can easily see the effect this would have on territories that are surrounded by water and for persons whose livelihoods are so closely connected to the environment (Niranjan, 2023). Due to their small size, remoteness, spatial isolation, and dependence on trade, SIDS are particularly vulnerable to shocks and stressors (UNDESA, 2022). When one considers the hardships associated with climate change, one can see how environmental migration can occur

T. Affonso (✉)
The University of the West Indies, St. Augustine, Trinidad and Tobago
e-mail: timothy.affonso@sta.uwi.edu

D. D. Joseph et al. (eds.), *Climate Change*,
https://doi.org/10.1007/978-3-031-83037-2_2

resulting in the phenomenon of physical displacement. This displacement, coupled with extreme weather phenomena caused by climate change, has a trickle-on effect on food security, job stability and access to fundamental human rights, such as the right to education and healthcare.

In addition to these existing challenges, there are the Sustainable Development Goals (SDGs) which seek to guide states in achieving targets directed at global development objectives. However, SIDS, which are already experiencing challenges because of climate change, encounter practical obstacles in simultaneously achieving the SDGs set by the international community. The foregoing presents the context for looking at the multi-faceted difficulties of the most vulnerable in the Caribbean region who are already facing various threats to their human existence in societies, even in the absence of SDGs and the persistent threat of climate change. While there is a widespread sensitivity to climate justice, there is a lack of consciousness of the disproportionate impact on already marginalized groups (Premkumar & Atanasova, 2023). It is this situation which will be examined in this paper, paying careful attention to the unique perspective of vulnerable groups set against the backdrop of the interaction between the plight of SIDS due to climate change and the obligations to achieve the SDGs. The first part of the paper will explore the impact of climate change on SIDS with a particular focus on the challenges which have been disproportionately experienced by vulnerable groups in society. The second part of the paper will explore the challenges faced by SIDS in achieving the SDGs and the burdens the SDGs place on SIDS' existing resources. The third part of the paper will critique existing support systems to help SIDS achieve the SDGS and there will also be recommendations which may allow for the SDGs to more realistically be achieved by SIDS in the Caribbean.

Climate Change and Caribbean Smalls Island Developing States (SIDS)

Direct Effect on the Environment

SIDS are characterized by certain common inherent characteristics (Sachs et al., 2021). They have minimal elevation above sea level and are directly affected by rising sea levels (Sachs et al., 2021). The close ties between the SIDS and the environment make climate change a threat to their very existence. The reality is that changes in weather patterns; increased sea levels; flooding and consequential impact on farming and housing are daily phenomena in the SIDS of the Caribbean. Of the Caribbean's 44 million inhabitants, more than fifty per cent live within 1.5 km of the coastline, with one-third of the population residing in low-elevation zones, less than 10 m above sea level (Mycoo, 2017). As of 2017, an estimated 22 million people in the Caribbean live below six metres of elevation (UWI, 2023).

Additionally, the evidence shows that the Caribbean chain of islands is the second most hazard-prone region in the world, due to its geographical location and the concentration of its population in coastal areas (Bleeker et al., 2021). Extreme weather events continue to be the main cause of disasters in the Caribbean and globally and they are magnified for Caribbean SIDS compared to countries in other regions (UNDRR, 2009). It is therefore possible to assert that Caribbean SIDS are consequently exposed to vulnerabilities that hinder their developmental progress by mere virtue of their geography (Sachs et al., 2023a, 2023b). This unique threat faced by SIDS explains why Article 4.8 of the United Nations Framework Convention on Climate Change (UNFCC) lists "small island countries; countries with low-lying areas and countries prone to natural disasters", as meriting particular consideration for assistance (UNFCC, 1994).

From an economic perspective, some SIDS rely more on services (e.g. the Bahamas, and Barbados), while others are more natural resource based (e.g. Trinidad and Tobago) (Sachs et al., 2023a, 2023b). Any changes to the weather directly affect the heart of regional economies and impact regional food security. This is because in many SIDS, local food production is challenged by the small size, the lack of arable land, the limited freshwater resources, the high vulnerability to natural disasters and the fragile natural environment which is exposed more and more to the impacts of overexploitation of oceanic resources, marine pollution and climate change (Massa, 2021). Furthermore, these sensitive small economies, which also depend on tourism, are directly affected by bleaching of coral reefs due to rising sea temperatures and an increase in extreme weather events (United Nations Climate Change, 2014) which further affects economic growth regionally.

There is even a threat to biological diversity as a result of climate change. New studies highlight large population reductions, with an extinction of one hundred per cent for endemic species within insular biodiversity hotspots such as the Caribbean (IPCC, 2022). A concerning reality is the drought risk projections in the Caribbean SIDS which indicate that a 1 °C increase in temperature could result in a 60% increase in the number of people projected to experience severe water resource stress from 2043 to 2071 (IPCC, 2022).

The foregoing data show that climate change threatens the very existence of Caribbean SIDS, impacting the movement of people, territorial loss, food insecurity and loss of biodiversity. These effects of the change directly influence survival of Caribbean people, the physical integrity of nation states and eradication of entire cultures and erodes the economic stability of the Caribbean region as a whole. The economic and financial cost of climate change will now be addressed from the perspective of Caribbean SIDS.

Consequences for Human and Economic Resources

The link between Caribbean economies and the environment is a sensitive one and climate change has directly affected the core agricultural products of the region. Unprecedented weather occurrences have decimated core exports of nutmeg in Grenada, sugar cane in Cuba and bananas in Jamaica, St. Lucia and Dominica (United Nations Climate Change, 2014). Additionally, as one source reveals, agricultural exports in the Caribbean region are "a sputtering engine of economic growth" in large measure due to the progressive decline of operations in the absence of specialized market access (UNECLAC, 2011).

This point raises an additional dimension to the sensitivity of Caribbean economies to weather changes, and that is the idea of global competitiveness of regional products. As small regional economies are forced to operate within national boundaries threatened by climate change, they just cannot compete with international economies of scale. The Caribbean's share of world merchandise exports has been declining steadily, falling from three per cent in 1970s to nearly a quarter per cent in 2012 (The World Bank, 2016). Even at an inter-regional level, the World Bank (2016) has stated that Caribbean maritime connectivity is relatively limited compared to other regions, as well as when comparted to other small island economies. The totality of these factors results in a narrowing of economic activity. The situation is only made worse when one considers that, in the Caribbean, the damage caused by climate-related and earth-related hazards is estimated at $12.6 billion per year (Akiwumi, 2022). With a reduction in economic activity, coupled with actual losses, the Caribbean SIDS find themselves in an unenviable position in the developmental race.

Of great concern is the fact that the overall population in the Caribbean, doubled in size in the latter half of the twentieth century, contributing to the rise of densely populated, informal settlements in [...] low-lying, vulnerable areas (Mycoo, 2017). These trends continue with projections that urbanization will increase threefold in Antigua and Barbuda, Barbados and the Bahamas by 2050 (Mycoo, 2017). Research has also shown that sea level rise will be between 1 and 2 metres above present levels by the end of the twenty-first century (Caribsave, 2010). From an almost fatalistic perspective, it has been advanced that sea level rise will continue for centuries after 2100, even if global temperatures are stabilized at 2 or 2.5 °C, which means that the current trend is a chronic and unidirectional, negative threat to coastal areas in the Caribbean globally (Caribsave, 2010). It can also be described as environmental erasure as states will permanently lose territory in a manner like the fabled Lost City of Atlantis.

The connection between climate change and the impact of regional resources is real and significant in terms of land loss, agricultural impacts, population displacement and erosion. Further, these impacts only worsen as time passes. Based on forecasts, it has been predicted that total dryland losses can reach $21.3 billion in 2050 and as much as $73.9 billion losses in 2080 (Caribsave, 2010) which will force internal movement of populations and mass emigration. In fact, four Caribbean countries (Puerto Rico, Haiti, the Bahamas, Dominica) were among the top 15 countries most affected by extreme weather-related events in the world between 2000 and 2019 (Migration Data Portal, 2021). Also, between 2010 and 2020, there were a total of 9.14 million new disaster-related displacements recorded in the Caribbean, and a stock of almost 100,000 remained displaced by the end of 2020 (IDMC, 2021).

In the light of the existing and future threats posed by climate change on islands of the Caribbean, it is necessary to discuss the concept of adaptation. Adaptation is a process of adjustment to actual or expected climate and its effects (Robinson, 2020). When overlaid with other environment and developmental challenges such as pollution, crime and corruption in a small island context, it requires urgent physical, political, economic, social and institutional adjustments of an incremental and transformational nature in order for countries to identify sustainable pathways (Nurse et al., 2014). Unlike land-locked states in the temperate zone, which do not tangibly witness the increasing sea levels, the Caribbean SIDS must prioritize climate adaptation in order to survive. This prioritization will take place at national levels across the region and will limit the already stretched resources of Caribbean governments.

The combined impact of climate change on physical and human resources not only has realized the negative potential on regional economies but also has a more devastating impact on the ability of Caribbean people to thrive. It is this looming reality that brings to focus human rights-related issues associated with climate change, particularly for Caribbean SIDS.

Human Rights Impact

The effect of global boiling on the Caribbean has immediate human rights implications for nationals. It is a delicate issue which crosscuts themes of population expansion; loss of physical territory of states; food security concerns; and national resource limitations. Even in the absence of any of these factors, vulnerable groups in Caribbean societies find themselves on the fringe of society and struggle to access basic fundamental rights. The challenges which are attendant to these vulnerable groups in society, such as women, children, persons of colour and sexually diverse persons, are only exacerbated by the effects of climate change.

It must be noted that "climate change exacerbates the pre-existing inequalities found in society such as housing and medical care, among many others, leading to [vulnerable groups] being disproportionately affected during climate disasters by the wider effects of climate breakdown on society" (Premkumar et al., 2023). Therefore, what can be seen from this discussion is that climate change directly affects SIDS and the consequence of this relationship is a burden to the already disadvantaged states of the Caribbean. While climate change threatens, at the macro level, the very existence of Caribbean SIDS, vulnerable groups within these societies encounter even greater obstacles to their survival.

When states, with limited resources, must help their populations during natural disasters or continuously provide support to the national community based on the relentless effects of climate boiling, vulnerable groups struggle to live. These uniquely threatened groups, which include women, children, racial and ethnic minorities throughout the Caribbean SIDS, continuously suffer from gender-based violence and even state-sanctioned discrimination. Adding climate change to that lived experience of these vulnerable groups and a need for governments to re-allocate limited resources, there is a tendency to further subject the societal outliers to a place of invisibility.

The impact of climate change on the spectrum of vulnerabilities in society is devastating. The natural consequences of climate change and global boiling for Caribbean SIDS can lead to food insecurity, drought, air pollution and flooding (APHA, 2023). It is therefore necessary to show how the UN SDGs for Caribbean SIDS place a further burden on these resource-limited states, thereby making the reality for minority groups in the face of climate change an almost insurmountable obstacle. Women, children, older adults and persons with disabilities are more vulnerable to health stressors such as extreme heat, floods and poor air quality (Malina, 2009). Data from the Caribbean show that women are more at risk of the impacts of climate change than men (UN Women, 2022). This phenomenon is because of the structural barriers to equality based on harmful gender stereotypes and inequality of power in decision-making that increases women's vulnerability and reduces their opportunities and access to build resiliency (UN Women, 2022).

Adding to this discussion is the fact that more than 400,000 children in the Caribbean SIDS were displaced by hurricanes in 2017 alone (UNICEF, 2019). Extreme weather phenomena increase due to climate change and UNICEF warned that without urgent action to mitigate the effects of climate change, the increasing proportion of severe storms would likely result in similarly high levels of forced displacement in the coming decades (UNICEF, 2019). Further, displaced children might be left with limited or no access to the essential services they need to thrive including education, protection and health care (UNICEF, 2019). There are also real challenges faced by sexually diverse people as a result of climatological phenomena. The Lesbian, Gay, Bisexual, Transgender and Intersex (LGBTI) community and non-binary individuals are uniquely vulnerable to disasters and climate change and poor displacement and migration outcomes, due to their societal exclusion and discrimination (Bleeker et al., 2020).

The totality of the present discussion demonstrates that SIDS are uniquely prone to the negative side effects of global boiling. The consequences of the boiling of the earth bring the existence of the entire Caribbean region to the fore. The reality for the region is a fight for survival which, at best, appears to be more akin to a slow delayed death. States which face resource challenges and threats to human survival have a practical obstacle in achieving globally set standards for development. The discussion of Caribbean SIDS achieving SDGS will now be addressed.

Challenges of Sustainable Development Goals (SDGs) for SIDS

The SDGs are meant to operate as universal targets for the international community in achieving global development. The 2030 Agenda for Sustainable Development refers in paragraph 23 to the empowerment of the most vulnerable and includes, "all children, youth, persons with disabilities, older persons, refugees, people living with HIV/AIDS" (UNSDG, 2016). However, there are systemic challenges to the realization of that objective because the capacity of all states is not equal. To this end, some have argued that the SDG framework hides multiple axes of inequality, creates fallacious claims of success and closes the window for potential improvement (Choudhary, 2023). Even in situations where the international community seeks to aid less developed nations so that they may realize the SDGs, there has been criticism that such aid finds its roots in the systems of oppression that perpetuate racist, colonial and sexist relations (Clements & Sweetman, 2020). Some have even argued that the SDGs are really an issue of international fairness (Barthelemy, 2016). Fairness can be explained from a perspective of whether the SDGs, as developed, are realistically capable of being achieved by all states. The issue of fairness penetrates the resolution to this issue because it demands an investigation into the underlying rationale for the framing of the SDGs in terms which are unattainable for most of the world's developing nations. Recognizing that international policy development is closely tied to global economic power, the SDGs do not tackle the root causes of inequalities and lack the potential to meaningfully change the vectors of oppression and dependence into channels of equity and genuine collaboration (Open Edition, 2023). Some have even argued that the SDGs concretize economic and social disparities among states (Blicharska et al., 2021).

From a human rights perspective, rendering the most vulnerable in society voiceless in global development excludes solutions which have the potential to address their challenges. In essence, SDGs have the potential to perpetuate inequality and further subjugate vulnerable groups to majoritarian standards. In so doing, their vulnerability is entrenched and the SDGs become a tool for reinforcing societal stratification. This could explain why Clements and Sweetman (2020) blame the SDGs for not questioning the unequal power relations at the root of the issues that the SDGs claim to tackle. There have been practical challenges to the creation of targets which meaningfully address vulnerabilities because the disaggregated data needed to tackle all vulnerable groups (children, persons with disabilities, older persons, indigenous peoples, refugees) are sparse (United Nations Stats, 2016). Based on the 2016 UN Report on the SDGs, the benefits of development are not equally shared and few of the current indicators are able to shed light on the situations of migrants, refugees, older persons, persons with disabilities, minorities and indigenous peoples (United Nations Stats, 2016).

Recognizing the inherent flaws of the SDGs, when they are imposed on Caribbean SIDS, they create a metaphoric storm. SIDS face a unique set of vulnerabilities which impede their ability to achieve sustainable development (Sachs et al., 2023a, 2023b). Fragile ecosystems, high vulnerability to climate change, natural disasters, external economic shocks, distance from global markets, and high rates of all forms of malnutrition and food-related diseases are just some of the challenges hindering the development process for SIDS (Food & Drug Organisation, 2023). While the foregoing may make the idea of sustainability and Caribbean SIDS appear to be incongruous, the reality is that the close ties between these states and their environment, coupled with their limited resources, make them much more innately sustainable than most other states. There is an unfortunate fact due to their small size and unique vulnerability to environmental changes. Any minor adjustment in the delicate relationship between the environment and SIDS could represent catastrophic tragedy at a national and regional level. There are therefore disproportionate obstacles for SIDS in achieving the SDGs than for other members of the international community. Even if that were not the case, the conditions of the most vulnerable in the Caribbean SIDS would still not benefit from any targeted goal to improve their lives.

The reality is that achieving the SDGs has a direct economic impact for states. Development occurs at a very high cost. The United Nations Conference on Trade and Development (UNCTAD) averaged the cost to achieve any of the six transformative "pathways" to sustainable development to be between $5.4 trillion USD and $6.4 trillion USD annually from 2023 to 2030 which translates to between $1179 USD and $1383 USD per person each year (2023). When expanded to cover all developing economies, using the median per capita cost for the 48 economies in the study, the total annual needs fall between $6.9 trillion USD and $7.6 trillion USD (UNCTAD, 2023). This is far more than the financial capabilities of the region and was reinforced by the 2023 United Nations Report on the achievement of the SDGs. Based on this resource limitation, some writers have advocated for international partnership or "global coupling" which allows stronger states to support weaker states to achieve the SDGs (Blicharska et al., 2021). This bridging of the economic gap was further developed by Filho et al., in their proposal for sustainable financing options (2022). However, it must be reiterated that if the achievement of the SDGs involves the imposing of debt on SIDS and the developing world, then global development is not meaningful. With the unique positioning of Caribbean SIDS in the international economic landscape, there is a practical limitation of available financing. Debt financing cannot be the way forward.

SIDS: Practical Realization of SDGs and Confronting Climate Change

It is commonplace that issues of development are relegated to an issue of financing. The countries with more substantial economic resources operate in a place that is limited only by the pace at which developmental initiatives can be realized. However, development, at a sustainable level, is not at the top of the priority listing for developing countries much less, SIDS of the Caribbean. That is not to say that national development is not an objective but that it must be balanced with many other pressing national concerns such as employment, health care, education, due process set in a national context and not an international comparative one.

Therefore, of particular relevance to the present discussion is that the failure of the SDGs to adapt to the different resources of SIDS in their universality of scope extends also to its failure to specifically identify action to protect the most vulnerable groups in society. It is in this context that the SDGs present no realistic hope for vulnerable groups in SIDS, even if the targets are met because they were never directly considered at the inception. As stated by Guerson, Morsink and Munoz (2023) "the Caribbean is the most exposed region to climate-related natural disasters with estimated adaptation investment needs of more than $100 billion, equal to one-third of its annual economic output". However, Caribbean countries have only been approved for about $800 million from climate funds such as Green Climate Fund, Global Environment Fund and Adaptation Fund (Guerson et al., 2023). It is consequently understandable that in the overall ranking of 193 UN Member States' achievement of the SDGs by 2023, six islands of the Caribbean were not ranked and the remaining English-speaking Caribbean states ranked between 82 (Jamaica) and 124 (The Bahamas) (Sustainable Development Report, 2023). It is essential to note that the ranking is based on a score that measures the total progress towards achieving all 17 SDGs and even the Jamaican score is only 69.59/100 (Sustainable Development Report, 2023). The Jamaican performance is commendable but even with that performance, there are concerns about the region underperforming at that international level and whether, with the precipitated climate change challenge, there will be a further disadvantage to the states in achieving the SDGs.

Based on the Special Edition of the Sustainable Development Goals Report of the United Nations, the earth is "standing at the brink of climate calamity" (UN DESA, 2023). With specific reference to the unmet needs of the developing world in achieving the SDGs, the Report shows that developing countries require nearly $6 trillion USD by 2030 in order to meet the SDGs. Additionally, highly vulnerable regions experience fifteen times higher mortality rates from disasters compared to very low vulnerability regions. Although the report seeks to offer guidance in simple terms on how to achieve each SDG, the resources necessary to realize these goals are not identified. This gap in the Report leaves the most vulnerable nations in an unenviable position of global neglect.

The only way in which Caribbean SIDS can be realistically expected to achieve the SDGs at a pace of more developed nations is to ensure that the necessary resources are provided to the states. The mechanism for this must be urgently addressed because for the countries of the Caribbean region, global boiling is not an inconvenience but a threat to its very survival. Leaving this region behind in the achievement of the SDGs is tantamount to a facilitated annihilation of an entire geographical region because inaction or action at a pace that does not protect the region will see the decimation of the home and people of the Caribbean.

Therefore, what needs to be done at this stage is the monetizing of purpose by the international community. Beyond the preparation of rankings and reports, the data must feed into real action that can result in meaningful change. If that is not the rationale for the work of the international community and organizations, then the SDGs are just another way in which developed countries can celebrate their achievements in addressing concerns caused by them and which affect others. Suggestions by the International Monetary Fund for private funding, loan arrangements and green initiatives will only serve to further plummet regional economies into debt which will have the natural consequence of states choosing survival or poverty. This, it is suggested, cannot be the way forward.

There needs to be immediate action taken by the developed world as the precipitation of climatological deterioration falls at the feet of industrialized societies, who have for centuries profited from environmentally deleterious economic activity. There is research that seeks to attribute historical responsibility for climate change by looking at the cumulative amount of carbon dioxide (CO_2) emitted since the start of the industrial revolution. In ranking countries on this scale, the United States of America assumes first place with more than 509GTCO_2 since 1950 and is responsible for the largest share of historical emissions (20% of total emissions) (Evans, 2021). The next rankings are China with 11%, Russia with 7%, Brazil with 5% and Indonesia with 4% (Evans, 2021). It should, however, be noted that taking a per capita emission perspective in 2021, the SIDS of Trinidad and Tobago ranked at number 5 based on its oil and gas industry, which then feeds into a sizeable national chemicals sector (Evans, 2021). That aside, it must be highlighted that small countries with high per capita emissions, like Trinidad and Tobago, are still relatively unimportant for warming overall (Evans, 2021). Additionally, with almost two hundred years of economic activity which profited the developed world and which negatively affected the environment, there is a duty, in terms similar to reparation, on the part of the developed world. It would be interesting to know the real environmental impact of Caribbean SIDS not achieving the SDGs as compared to the shortfall of the more successful developed world.

This duty is grounded in the benefit which accrued to the developed nations and should factor into a reformed system of climate justice. The need for funding in the achievement of SDGs ought not to be a choice but a duty on the part of the international community as it relates to global development and environmental protection. Any discussion on development which ignores the disproportionate duty of the developed world and the impact on the developing world is destined to fail. The reality is that this discussion may already be too late to undo the harsh effects of climate change on the world as we know it. However, politicizing the discussion and casting blame offers little real hope for change. It is ironic that the most vulnerable economies (SIDS) who have had little impact on the climate change phenomenon are so intimately affected by its consequences. However, SIDS' action will do little to effect change and the discourse must therefore change from how to apportion shared responsibility to the demand for action by the wealthiest states in the world.

Recommendations

The creation of any policy for Caribbean SIDS requires careful tailoring with as much attention placed on form as on substance. Moving forward requires urgent, deliberate and efficacious action. Linked to the complex and multi-dimensional challenges which face the Caribbean SIDS in achieving the SDGs, there is a need for a multi-pronged policy tool of development which is created by a cross-section of stakeholders nationally and regionally so that priority is given to those states which face the most imminent threat and/or risk based on national vulnerability to environmental degradation. It is important to have the voices of those who are directly affected by climate change involved in the creation of a policy that seeks to find a solution to their unique problems. This would, of necessity, include inter alia, farmers, youth, women, business groups, government and elderly. Ultimately, the developmental policy tool must approach environmental protection and economic stability simultaneously. It is in birthing a policy of this nature, through this process, that the macro and micro can be engaged in a meaningful way. This will ensure that individuals and nation can be protected. Furthermore, when the environment and the economy are targeted, there will naturally be a requirement to safeguard the national and regional resources which include the human resource capacity. It results organically in the protection of those who suffer under multiple axes of economic, social and geographic oppression. Of specific importance is a modification of the international refugee framework so that it includes climate migration as a recognizable form of refugees and the removal of the persecution requirement in the existing international system. There is also a need to include in the developmental tool space for national water policies and national and regional equity policies to ensure that no one and no state is left behind.

Drawing on the undergirding theme of inadequate financial resources for the achievement of the SDGs by SIDS, the developmental policy tool must guarantee aid for the at-risk economies of SIDS. Recognizing the historical role and benefit gained by the deleterious conduct of states from the industrial revolution to date, there needs to be either a global coupling or economic contributions by developed states. These funds should be paid into a pool of funds which is managed by regional bodies and accessed without the bureaucratic process inherent in financial aid in the UN system. While some may argue that the developed states owe no responsibility to the weakest vulnerable states globally, the reality is that an adoption of that approach to supporting states achieving the SDGs reveals the disingenuous nature of the SDG narrative. This reinforces a possible view that the SDGs were never meant to be a realistic target of global development but a system which allows the political and economic global leaders to re-assert their dominance and perpetuate inequity through the pretence of international peace and security.

Conclusion

The global community requires urgent collective and collaborative action to stem the tide of climate change. The effect of climate change is undeniable but it is felt more directly by some states than others. While it is undeniable that global boiling affects us all, the very existence of some states is dependent on the stemming of the boiling of the earth. Caribbean SIDS have a direct connection to the environment in which they are geographically located. With a significant proportion of the region living in coastal zones, climate change and global boiling threatens the very existence of the entire nations. The difficulties faced by Caribbean SIDS because of climate change are only further exacerbated by the international standards set by the SDGs. This is because the SDGs, while they aim to realize a universal standard of development, do not consider the significant capacity limitations of the Caribbean SIDS. These limitations are greatly impacted by natural disasters such as flooding, drought, rising water levels and human displacement. There is consequently a close relationship between climate change and SIDS.

Adding another layer of challenges to this discussion is the fact that within Caribbean societies, there is a vulnerable group which does not even have legal protection by regional governments. Vulnerable groups exist on the periphery of society and do not benefit from the same substantive rights as other individuals. When their challenges are put in the context of nations with already stretched resources, considerations such as climate change and SDGs plummet the priority of these groups. This results in a modern construct of state-sanctioned neglect and discrimination. Whenever there are threats to national development, those most affected are the most vulnerable in society.

It is therefore of urgent importance that the wealthier states, who have benefitted from economic activity of such a nature that the environment was negatively affected, must bear the financial burden of addressing the climate change phenomenon. Any discussion of climate financing in the form of debt for SIDS is an untenable proposal and disregards the situation in which we, as a people, find ourselves. We must guard against an "othering" of the plight to climate change. It is not a north–south discussion. The urgency of the need for action has passed and we are now in a state of predicting what happens next as a result of a failure to act fast enough. The harsh reality is that the SIDS of the Caribbean are objects in a discourse that directly affects their very existence but they are powerless to meaningfully redirect the course of climate change. The SIDS must therefore hope that the developed world will put global environmental protection above individual economic protection of status. Until individual self-preservation is abandoned in the climate change fight, we will all collectively lose.

References

Akiwumi, P. (2022). Climate Finance for SIDS is shockingly low: Why this needs to change. https://unctad.org/news/blog-climate-finance-sids-shockingly-low-why-needs-change

American Public Health Association (APHA). (2023). Climate changes health: Vulnerable populations. https://www.apha.org/topics-and-issues/climate-health-and-equity/vulnerable-populations

Barthelemy, P. (2016). Achieving SDGs: An issue of fairness towards least developed countries. https://www.iddri.org/en/publications-and-events/blog-post/achieving-sdgs-issue-fairness-towards-least-developed-countries. Accessed July 19, 2024.

Bleeker, A., Escribando, P., Gonzales, C., Liberati, C., & Mawby, B. (2021). *Advancing gender equality in environmental migration and disaster displacement in the Caribbean*. United Nations Publication.

Blicharska, M., Teutschbein, C., & Smithers, R. J. (2021). SDG partnerships may perpetuate the global North-South divide. *Open Scientific Reports, 11*, 22092.

Choudhary, N. (2023). Critiquing the SDG framework through the lens of goal two: Empirical reflections from two case studies in India. *Forum for Development Studies, 50*(2).

Clements, M. A., & Sweetman, C. (2020). Introduction: Reimagining international development. *Gender & Development, 28*(1), 1–9.

Evans, S. (2021). Analysis: Which countries are historically responsible for climate change? https://www.carbonbrief.org/analysis-which-countries-are-historicallyresponsible-for-climate-change/. Accessed May 10, 2024.

Filho, W., Dinis, M., Ruiz de, M. S., Doni, F., Eustachio, J., Swart, J., & Paco, A. (2022). The economics of the UN sustainable development goals: Does sustainability make financial sense? *Discover Sustain, 3*(20). https://doi.org/10.1007/s43621-022-00088-5

Food and Drug Organisation. (2023). Boosting small island developing states' ability to achieve the sustainable development goals. https://reliefweb.int/report/world/boosting-small-island-developing-states-ability-achieve-sustainable-development-goals. Accessed April 18, 2024.

Guerson, A., Morsink, J., & Munoz, S. (2023). Caribbean climate crisis demands urgent action by governments and investors. https://www.imf.org/en/Blogs/Articles/2023/06/27/caribbean-climate-crisis-demands-urgent-action-by-governments-and-investors

Internal Displacement Monitoring Centre (IDMC). (2021). Global internal displacement database. https://reliefweb.int/report/world/internal-displacement-index-2021-report?gad_source=1&gclid=CjwKCAjw88yxBhBWEiwA7cm6pZ2j7rLFN9JvK8rtfea-VNO8KoSZ-0Ck_zg8xGMqojvvQrxkDe6nNxoCtZoQAvD_BwE. Accessed December 15, 2023.

Intergovernmental Panel on Climate Change (IPCC). (2022). Climate change 2022: Impacts, adaptation and vulnerability. https://reliefweb.int/report/world/climate-change-2022-impacts-adaptation-and-vulnerability?gad_source=1&gclid=CjwKCAjw88yxBhBWEiwA7cm6paL3gWMI8lSxVgzuAM4sgwBPJUpOuLuERi8GLQOpOmn9z1ZwPM9fgRoCFOIQAvD_BwE

Malina, B. (2009). Identifying vulnerable subpopulations for climate change health effects in the United States. *Journal of Occupational and Environmental Medicine, 51*(1): 33–7.

Massa, I. (2021). More vulnerable SIDS face more difficulties to achieve the SDG 2—zero hunger. *Sustainable Development Solutions Network.*

Migration Data Portal. (2021). *Migration data in the Caribbean.* International Organization for Migration.

Mycoo, M. (2017). Beyond 1.5°C: Vulnerabilities and adaptation strategies for Caribbean small island developing states. *Reg Environmental Change, 18*, 2431–2353.

Niranjan, A. (2023). Era of global boiling has arrived. The Guardian. https://www.theguardian.com/science/2023/jul/27/scientists-july-world-hottest-month-record-climate-temperatures#:~:text=The%20era%20of%20global%20warming,the%20beginning%2C%E2%80%9D%20Guterres%20said. Accessed July 27, 2024.

Nurse, L. A., R. F., McLean, J., Agard, L. P. B., Duvat-Magnan, V., Pelesikoti, N., Tompkins, E., & Webb, A. (2014). *Small islands*. In V. R. Barros, D.J. Field, M. D. Dokken, K. J. M. Mastrandrea, T. E. Bilir, M. Chatterjee, K. L. Ebi, G. Y. O. R. C. Estrada, B. Girma, E. S. Kissel, A. N. Levy, S. MacCracken, P. R. Mastrandrea, & L. L. White (Eds.). *Climate Change 2014: Impacts, Adaptation, and Vulnerability. Part B: Regional Aspects. Contribution of Working Group II to the Fifth Assessment Report of the Intergovernmental Panel on Climate Change*. Cambridge University Press.

Premkumar, S., & Atanasova, L. (2023). The disproportionate impact of climate crisis on the LGBTQIA2S+ community. https://www.greenpeace.org/international/story/60078/impact-climate-crisis-lgbtqia2s-pride-month/#:~:text=Furthermore%2C%20climate%20change%20exacerbates%20the,of%20climate%20breakdown%20on%20society. Accessed June 1, 2024.

Robinson, S. (2020). Climate change adaptation in SIDS: A systemic review of the literature pre and post the IPCC fifth assessment report. *WIREs Climate Change, 11*(4).

Sachs, J., Massa, I., Marinescu, S., & Lafortune, G. (2023). The decade of action and small island developing states: Measuring and addressing SIDS' vulnerabilities to accelerate SDG progress. https://jamaica.un.org/en/137561-decade-action-and-small-island-developing-states-measuring-and-addressing-sids%E2%80%99. Accessed July 27, 2024.

United Nations Department of Economic and Social Affairs (UN DESA). (2023). The sustainable development goals report 2023: Special Edition. https://reliefweb.int/report/world/sustainable-development-goals-report-2023-special-edition?gad_source=1&gclid=CjwKCAjw88yxBhBWEiwA7cm6pW2vhoqNbtAUOON264FPCIWJ2AwFvheYoZWS0SoVlHubbeBqCRfm8BoCUc0QAvD_BwE

Sachs, J., Kroll, C., Lafortune, G., Fuller, G., Woelm, F. (2021). *The Decade of Action for the Sustainable Development Goals: Sustainable Development Report 2021*. Cambridge University Press.

Sachs, J. D., Lafortune, G., Fuller, G., & Drumm, E. (2023). Implementing the SDG Stimulus. Cited in Sustainable Development Report, "Ranking—The Overall Performance of all 193 UN Member States". https://dashboards.sdgindex.org/rankings

The World Bank. (2016). Trade matters: New opportunities for the Caribbean. Caribbean Growth Forum. https://documents1.worldbank.org/curated/en/738061467998816272/pdf/93184-v2-P146683-Box391466B-PUBLIC-trade-caribbean-web.pdf. Accessed July 8, 2024.
UNCTAD. (2023). UNCTAD Counts the costs of achieving sustainable development goals. https://unctad.org/news/unctad-counts-costs-achieving-sustainable-development-goals. Accessed September 18, 2024
UNDESA, UNDRR, AOSIS. (2022). Small island developing states (SIDS)—gaps, challenges and constraints in means of implementing the Sendai framework for disaster risk reduction. UNDESA, UNDRR, AOSIS.
UNECLAC. (2011). *The Economics of Climate Change in the Caribbean: Summary Report 2011.* UNECLAC.
UNFCC. (1994). Entered into force on March 21, 1994.
UNICEF. (2019). As impact of climate crisis worsens. https://www.unicef.org/jamaica/press-releases/impact-climate-crisis-worsens-caribbean-islands-see-six-fold-increase-estimated#:~:text=NEW%20YORK%2FPANAMA%2C%206%20December,new%20UNICEF%20report%20said%20today
United Nations Climate Change. (2014). Climate change impacts Caribbean economy. https://unfccc.int/news/climate-change-impacts-caribbean-economy
United Nations Office for Disaster Risk Reduction (UNDRR). (2009). *Global Assessment Report on Disaster Risk Reduction.* UNDRR.
UNSDG. (2016). Ensuring that no one is left behind: Reaching the most vulnerable. https://hlpf.un.org/2016/programme/ensuring-no-one-left-behind-reaching-most-vulnerable
United Nations Stats. (2016). The sustainable development goals report. https://reliefweb.int/report/world/sustainable-development-goals-report-2016?gad_source=1&gclid=CjwKCAjw88yxBhBWEiwA7cm6pRKWlZwFAixwZWYMFbyiYd01S6E2L340f0bN4vcD2aNT7v6WwgpmkxoCIxAQAvD_BwE
UN Women (Caribbean). (2022). Climate change and disaster risk resilience: Gender matters!. https://caribbean.unwomen.org/en/stories/news/2022/03/climate-change-and-disaster-risk-resilience-gender-matters
UWI. (2023). Caribbean under threat!—Part 2: 10 urgent takeaways for the caribbean from the UN IPCC's latest report on climate change. https://www.mona.uwi.edu/news/caribbean-under-threat-%E2%80%93-part-210-urgent-takeaways-caribbean-un-ipcc%E2%80%99s-latest-report-climate#:~:text=Among%20these%20include%20loss%20of,surpassed%20between%20now%20and%202040

CHAPTER 3

The Psychosocial Impact of Climate Change: A Recent Overview

Gervan J. Arneaud

Global climate change is increasingly becoming a problem for human well-being with immediate challenges to physical health and socioeconomic well-being. Rising global weather temperatures, an increased prevalence of natural disasters and diseases have caused significant psychosocial and psychological challenges globally (World Health Organization [WHO], 2023). The prevalence of mental health disorders has increased exponentially due to extreme weather conditions, whether it is due to directly experiencing a weather-related catastrophe or vicariously through the accounts of others. For example, rising temperatures, one of the most common features of climate change, have been linked to increases in the rate of suicides (Burke et al., 2018; Cheng et al., 2021; Williams et al., 2016). Another example is the increased incidence of post-traumatic stress disorder in individuals who experienced climate events like hurricanes and flooding (Chen & Liu, 2015; Golitaleb et al., 2022).

G. J. Arneaud (✉)
Journey Behavioral Health Clinic, Trinidad and Tobago, Trinidad and Tobago
e-mail: gervan@journeybehavior.com

D. D. Joseph et al. (eds.), *Climate Change*,
https://doi.org/10.1007/978-3-031-83037-2_3

There are also larger-scale social and community psychosocial impacts as individuals may have their homes destroyed and need to relocate to new communities, regions, or even countries. Many countries, such as Brazil, Mexico, and the United States, also face socioeconomic challenges through damage to crops affecting food production, food security, import revenue, damage to and loss of homes and infrastructure, increasing prevalence of health problems affecting employment, and loss of tourism revenue (Bucci et al., 2023; Gay et al., 2006; Pham et al., 2019; Santos et al., 2018). Therefore, it is necessary to address the psychosocial challenges of global climate change to enhance our ability to safeguard well-being amidst climate-related threats and ensure a healthier future for individuals and communities. This chapter will review recent research in climate and behavioural sciences on the psychosocial impacts of climate change and interventions to address them.

Brief Overview of Climate Change

Climate change refers to the global or regional shifts in temperature and weather patterns over an extremely long period, often hundreds or thousands of years (United Nations, n.d.a). Climate change is a natural process as the climate has changed many times over millions of years because of the Earth's natural processes such as volcanic eruptions (Boudreau et al., 2022). However, since the mid-eighteenth century, climate change has been advancing much more quickly due to human activity such as the popularisation of burning fossil fuels like coal, oil, and gas for energy for manufacturing, generating electricity to power homes, other buildings, and various modes of transport (Intergovernmental Panel on Climate Change [IPCC], 2021). This type of climate change is referred to as anthropogenic climate change and happens at a much faster rate than if the process were to occur without human intervention. According to the IPCC Sixth Assessment Report (IPCC, 2023), the global surface temperature rose by 1.59 °C between the mid-nineteenth century and 2020, indicating a shift in the planet's ecosystem. This shift has implications for more extreme weather events, melting polar ice caps and rising levels, disruptions in ecosystems and biodiversity loss, and risks to human health (WHO, 2015, 2023). The reliance on fossil fuels for the past few centuries has led to a drastic increase in greenhouse gas emissions such as carbon dioxide and methane. These greenhouse gases are so-called due to their ability to collect in the atmosphere and blanket the Earth trapping heat known as the greenhouse effect (United States National Aeronautics and Space Administration, n.d.). Greenhouse gases, therefore, act like the insulating glass of a greenhouse which allows sunlight into the atmosphere but traps the heat being reflected by the earth's surface. This leads to the subsequent increase in global temperatures which quickens the rate of climate change.

The use of fossil fuels to generate power is not the only cause of climate change, deforestation in industries such as agriculture, transport, and infrastructure is also another major cause of increased greenhouse gas emissions. Trees and plants are important in managing the levels of carbon dioxide in the atmosphere through photosynthesis where they convert carbon dioxide into oxygen and release it back into the atmosphere (National Oceanic and Atmospheric Administration, 2019). Plants are also a key part of the carbon cycle because of their ability to absorb and store carbon dioxide. When forests and vegetation are cleared, they release their stored carbon dioxide and cannot photosynthesise to manage the levels of carbon dioxide in the atmosphere. It is estimated that approximately twelve million hectares of forest are destroyed each year (UN, n.d.b). As a result, deforestation is responsible for roughly a quarter of global greenhouse gas emissions.

Characteristics of Climate Change

The term climate change is often used in reference of rising global temperatures, but the effects of climate change are much more widespread. First, the gradual warming of the climate results in changes in weather patterns such as more frequent droughts. For example, in 2021 the United Kingdom experienced its second wettest May since 1836, with Wales setting a new observed record approximately 30 mm higher than the previous record (Christidis & Stott, 2022). Conversely, in states like California and Nevada in the United States experienced record droughts with California's levels of precipitation fell to record levels for the first time since 1895, while Nevada experienced their second-highest temperatures (Hoell et al., 2022).

In addition to changes in weather patterns, other major features of climate change are the rising of ocean temperatures and increased frequency of storms (UN, n.d.b). As global surface temperatures increase, so too do ocean temperatures albeit at a much slower rate. However, despite the slower rate of warming than the surface, the rate of warming of the oceans has increased in the last couple of decades (UN, n.d.b). As the oceans warm, water expands which causes a rise in sea levels. Increased ocean temperatures also result in more evaporation which provides ideal conditions for increased rainfall and flooding and the development of more severe storms (Knutson et al., 2021).

Another consequence of climate change is the loss of biodiversity (UN, n.d.b) as global temperatures increase, the environments that house many species of plants and animals begin to change and may no longer be able to support these forms of life. Cahill et al. (2013) summarised some of the ways in which climate change has the potential to contribute to the extinction of some species. These included abiotic factors like individual species' physiological ability to withstand changes in temperature and precipitation; and biotic factors such as negative impacts on beneficial species and positive impacts on harmful species, as well as changes in the temporal relationships between species. A popular example of the effects of climate change on biodiversity is the effect of global warming on the ice caps. As global temperatures increase, the Arctic ice caps melt which provides less habitat for the plants and animals that live there.

Impacts of Climate Change on Human Lives and Livelihoods

As the effects of anthropogenic climate change continue to worsen each year, so do the impacts on human lives and livelihoods. The effects of climate change have direct and indirect impacts on ecology, our ability to grow food, housing and infrastructure, which harms the health of humans and their well-being (UN, n.d.b).

Ecological Impacts

Climate change is often thought to be a primarily ecological issue with ecological consequences. Ecologically, marine life is particularly susceptible to the effects of climate change. For example, as the concentration of greenhouse gases increases in the atmosphere, the ocean will absorb more carbon dioxide. This reduces its pH making it more acidic through a process called ocean acidification (Mollica et al., 2018). Ocean acidification reduces the amount of carbonate ions available, therefore, affecting the ability of coral to construct their skeletons. Coral reefs are some of the most diverse ecosystems on Earth by providing food, shelter and substrate for a plethora of organisms, as well as significant cultural and economic value (Mollica et al., 2018). As coral reefs deteriorate, the loss of biodiversity and ecosystem can lead to reduced fish stocks, economic hardship for those who rely on them, as well as negatively impact the cultural identity of coastal towns and people.

Another more popular example is the impact of global warming on the polar bear population (Boonstra et al., 2020; Born, 2019). As global temperatures increase and the Arctic ice caps melt, polar bears lose the ice caps that function as their main habitat. This also affects seals which are polar bears' primary food source. Since seals rely on sea ice to rest, breed, and raise their young, warming temperatures and shrinking ice caps have the potential to force seal pups into the water before they are ready and able to survive (National Oceanic and Atmospheric Administration [NOAA] Fisheries, 2023). Additionally, as polar bears lose more and more of their habitat each year, they are driven further inland, therefore, increasing the likelihood that they will encounter humans and be injured or killed in self-defence.

Climate change significantly impacts marine and Arctic ecosystems, with repercussions for psychosocial well-being as biodiversity loss and habitat destruction intensify. Ocean acidification disrupts coral reefs, reducing fish stocks and affecting the economic and cultural stability of coastal communities. Similarly, polar bears face habitat loss due to Arctic ice melt, endangering their survival and affecting the species that rely on the same ecosystems. These environmental changes foster ecological grief and anxiety, impacting individuals and communities who depend on these ecosystems for livelihood, cultural identity, and a sense of place.

Economic Impacts

Climate change also has severe economic consequences through damage to property, infrastructure, and supply-chain demands (Pörtner et al., 2022). For example, in 2022, Hurricane Ian was the costliest hurricane in the history of Florida and the third most costly hurricane in the history of the United States (Bucci et al., 2023). Hurricane Ian was responsible for an estimated USD $112.9 billion worth of damage across the state of Florida. Ian produced a severe storm surge near southwest Florida where an estimated 900 structures were destroyed and 2200 damaged in the town of Fort Myers Beach alone. In the wider Lee County area, more than 5000 structures were destroyed and over 14,000 were severely damaged (Bucci et al., 2023). Hurricane Ian also affected the coastal areas of North Carolina and South Carolina and destroyed more than 100,000 homes on the island of Cuba. The economic devastation caused by climate events significantly impacts people's well-being by disrupting livelihoods, displacing families, and creating long-term financial strain that has implications on stress and anxiety.

Agriculture is another industry that is extremely vulnerable to climate change due to its large scale and reliance on weather patterns and environmental conditions. Rising temperatures and variations in precipitation have been shown to affect the growth and yield of crops (Malhi et al., 2021). For example, the Veracruz region in eastern Mexico is the second-largest coffee producer in the country and generates approximately 8% of Mexico's gross domestic product and employs 32% of the state's labour force (Gobierno del Estado de Veracruz, 2001, as cited in Gay et al., 2006). However, coffee production is extremely sensitive to weather and climate events such as droughts, floods, and heat waves which can significantly impact the final crop yield (Gay et al., 2006). The impact of climate change on coffee production is mostly negative as coffee yields decrease due to areas suitable for growing and cultivating coffee plants decreased and rising temperatures and humidity created favourable conditions for pests and diseases that negatively impacted coffee production (Pham et al., 2019). The vulnerability of agriculture to climate change affects people's well-being by threatening food security, employment, and economic stability, particularly in regions like Veracruz, Mexico, where coffee production is a major livelihood. As climate conditions become more erratic, crop yields decline, suitable farming areas shrink, and pests and diseases increase, intensifying financial and food-related anxieties for individuals and communities reliant on agriculture.

Climate change increasingly threatens physical and psychological well-being through direct and indirect impacts on human health. Severe weather events like hurricanes lead to direct fatalities and displacement, while climate-driven changes contribute to the spread of vector-borne diseases like Chagas and air pollution from dust storms, which increase respiratory issues. Additionally, extreme temperatures, drought, and other climate stressors have been associated with heightened psychological distress, worsened mental health, and increased suicide rates, underscoring the profound toll climate change takes on overall human health and mental well-being.

Impacts on Health and Well-being

As the impact of climate change continues to worsen societies' social, economic, and environmental systems, the effects on the physical and psychological well-being become more evident. Severe weather events like natural disasters have the potential to directly impact human health and well-being such as Hurricane Ian which was responsible for at least 156 deaths in the United States and 5 in Cuba. Of these deaths, 42% of deaths in the US and 60% of deaths in Cuba were considered "direct deaths," which is a death caused by hurricanes such as drowning in storm surge and freshwater flooding, as well as lighting and wind-related deaths (e.g. collapsing structures). There are indirect deaths that is caused by pollution, infectious disease, displacement, malnutrition, and psychological distress (UN, n.d.b; WHO, 2023).

An example of the role of climate change in illness and disease is the growing incidence of Chagas disease. Chagas disease is an infectious disease primarily caused by insects which house the virus known to cause Chagas disease. Acute Chagas disease can present without symptoms or with mild symptoms such as fever, fatigue, body aches, diarrhoea, and vomiting that make it hard to diagnose (WHO, 2024). If undiagnosed, the disease can progress to its chronic form which causes cardiac or gastrointestinal complications such as enlarged heart, oesophagus, or colon, heart failure, or cardiac arrest (WHO, 2024). Deforestation is a major cause of Chagas disease, as forests and trees are cleared away to make space for agriculture and livestock rearing has forced insects carrying the virus into high human populations (Coura, 2014; Santos et al., 2018; Sousa Júnior et al., 2017). In addition, the virus is also known to be carried by bats and rodents. The deforestation of the Amazon rainforest has caused an increase in vector-borne Chagas disease as 20% of the Amazon has been lost to deforestation since the 1990s.

In addition to infectious diseases, climate changes also have the potential to contribute to pollution which affects human health. One such example is dust storms. Dust storms refer to a distribution of small particles in the air typically from arid or semiarid regions such as the Sahara-Sahel region in northern Africa. These storms can travel thousands of kilometres from their areas of origin. Dust storms have been shown to be related to pulmonary health conditions like bacterial and fungal lung infections, asthma, and chronic obstructive pulmonary disease (Schweitzer et al., 2018). One study in Greece found that during dust storms, the number of presentations to the emergency department for dyspnoea or shortness of breath increased 3.6-fold (Lorentzou et al., 2019).

Climate change can also have the potential to impact one's psychological well-being. A recent review by Charlson et al. (2021) found that climate events such as heat, humidity, drought, wildfires, and flooding were associated with psychological distress, worsened mental health, increased psychiatric hospitalisations, higher mortality among people with pre-existing mental health conditions, and higher rates of suicide. Studies have shown that rising temperatures and humidity have been related to increased hospitalisations due to self-harm and suicide (Burke et al., 2018; Williams et al., 2016). Williams et al. (2016) reviewed data from the New Zealand Ministry of Health on discharges between 1993 and 2009 for self-harm and meteorological data from the United States National Institute of Weather and Atmospheric Research and found that a 1 °C rise in temperature was associated with a 0.7% increase in self-harm events that led to hospitalisation. Similarly, Burke et al. (2018) reviewed data on suicides in the United States from 1968 to 2004 and in Mexico from 1990 to 2010 and found a similar 0.7% increase in suicide rates in United States counties and a 2.1% increase in Mexican municipalities.

Climate change increasingly threatens physical and psychological well-being through direct and indirect impacts on human health. Severe weather events like hurricanes lead to direct fatalities and displacement, while climate-driven changes contribute to the spread of vector-borne diseases like Chagas and air pollution from dust storms, which increase respiratory issues. Additionally, extreme temperatures, drought, and other climate stressors have been associated with heightened psychological distress, worsened mental health, and increased suicide rates, underscoring the profound toll climate change takes on overall human health and mental well-being.

The Psychosocial Impact of Climate Change

According to the World Health Organization, anthropogenic climate change is a key threat to global mental health (WHO, 2022). Climate change is a complex issue with far-reaching effects that encompass not only environmental changes but also significant psychological and social changes. In the past couple of decades, researchers have begun to emphasise the role of the social sciences in preventing and dealing with the impacts of climate change. Doherty and Clayton (2011) categorised these psychological and social impacts of climate change into three main types: acute or direct impacts, vicarious or indirect impacts, and psychosocial impacts.

Acute or Direct Impacts

The acute or direct psychological impacts of climate change are the immediate effects on the prevalence or severity of mental health challenges among individuals and communities, including the implications for mental health services, and ongoing disruptions to the social, economic and environmental determinants that promote mental health and well-being (Doherty & Clayton, 2011). Examples of acute impacts are post-traumatic stress disorder, major depression, substance abuse, and higher rates of suicide following extreme weather events or natural disasters. An example of this is the prevalence of Post-Traumatic Stress Disorder (PTSD) following a flooding event. PTSD is a psychological condition that results from events where an individual believes there is a threat to their life or safety and experiences fear, terror, or helplessness (American Psychological Association, 2023). It is characterised by painful memories, flashbacks, avoiding activities or places related to the event, or physiological symptoms like exaggerated startle responses, disturbed sleep, or difficulty with concentrating or memory. PTSD is diagnosed using several criteria such as:

- Exposure to actual or threatened death or serious injury.
- The presence of intrusive symptoms associated with the event such as recurrent, involuntary, and intrusive memories or intense or prolonged psychological distress when exposed to cues that symbolise or resemble the traumatic event.
- Persistent avoidance of stimuli associated with the traumatic event following said event.
- Negative alterations in mood or cognition associated with the traumatic event following said event (American Psychiatric Association, 2013).

Floods are among the most severe and life-threatening natural disasters due to their unpredictable nature (e.g. flash flooding). Fast-moving floodwaters can generate large amounts of force that can easily sweep away people and debris and usually have long recovery periods due to infrastructure and material damage (Fontalba-Navas et al., 2017). Additionally, flooding can also impact on individuals' psychological well-being, sense of self, and sense of home and belonging as personal possessions like books, irreplaceable photographs, and marriage certificates are often destroyed (Carroll et al., 2009).

Chen and Liu (2015) reviewed 14 studies published between 2002 and 2012 that included scales or diagnostic criteria of PTSD following a flooding event and found a combined incidence of 15.74% or 1 in 6 people experienced PTSD. The prevalence of PTSD was determined by how recently the flooding event happened. For example, if the event occurred within the last 6 months, approximately 11% of people experienced PTSD. However, if the event occurred more than 6 months ago, the rate of PTSD increased to 16%. More recently, Golitaleb et al. (2022) reviewed 23 studies published between 2015 and 2020 and found a prevalence of PTSD at 29.48% or approximately 1 in 3 people. According to data from the World Mental Health Survey in 2016, the prevalence of PTSD among people exposed to a traumatic event was 5.6% (Koenen et al., 2017) indicating a significant difference between the estimated global prevalence of PTSD in individuals with a history of trauma and PTSD among individuals experiencing climate change-related events.

Another example of the acute impacts of climate change includes higher rates of suicide following climate events. Cianconi et al. (2020) reviewed 97 articles published between 1996 and 2019 on the effects of extreme weather (e.g. heat waves, flooding, hurricanes, drought) and psychiatric illness (e.g. depression, anxiety, PTSD, suicide). The findings were suicide rates that were influenced by extreme weather events such as rising temperatures and heat waves, tornadoes and hurricanes, drought. To further examine the relationship between rising temperatures and suicide, Cheng et al. (2021) conducted a longitudinal study that reviewed mortality data for intentional self-harm between 1999 and 2019 in the five most populated counties in California: Los Angeles County, San Diego County, Orange County, Riverside County, and San Bernardino County. They found that suicide rates increased by 0.82% for a 1 °C increase in temperature. These findings were slightly higher than earlier findings of a 0.7% increase in suicide rates from reviews in the United States, Mexico, and New Zealand (Burke et al., 2018; Williams et al., 2016).

Overall, the acute psychological impacts of climate change can be significant and manifest as immediate and severe mental health challenges among individuals and communities. Extreme weather events and natural disasters significantly contribute to conditions such as depression, PTSD, and increased rates of suicide and can have implications for mental health services, and ongoing disruptions to the social, economic, and environmental determinants that promote mental health and well-being.

Vicarious or Indirect Impacts

Doherty and Clayton (2011) described vicarious or indirect impacts of climate change as the intense emotional responses associated with individuals seeing and hearing about the effects of climate change around the world, as well as the anxiety and worry that individuals experience when considering current and future risks to humans and other species. These feelings have often come to be referred to as ecoanxiety which is a catch-all term that describes difficult or unpleasant feelings that occur because of the ecological crisis (Pihkala, 2020). Ecoanxiety itself is a multifaceted concept that has been widely discussed in a variety of fields such as social and political sciences, psychodynamics, psychosocial science, and climate science (Clayton et al., 2017, 2021; Lertzman, 2015; Materia, 2016).

Over the years different disciplines have attempted to define ecoanxiety. Within the psychological sciences, the American Psychological Association (Clayton et al., 2017, 2021) defined ecoanxiety as "a chronic fear of environmental doom" (p. 68). This was in relation to feelings such as helplessness, depression, fear, and resignation that individuals experienced because of gradual, long-term changes in the climate. This was particularly related to feelings of helplessness in making a difference against climate change. Additionally, psychodynamic research used the term 'environmental melancholia' to describe feelings of unresolved mourning in response to climate change (Lertzman, 2015) which mirrors ecoanxiety very closely.

Within the field of anxiety studies, Materia (2016) described the concept of climate state anxiety which refers to "the lived experience of anxiety as an expression of worry, concern, apprehension, unease and fear, brought about by changes in nature" (p. 32). Once again, this concept shares similarities in the description of ecoanxiety. There is a distinction between a form of state anxiety and a form of trait anxiety, a state anxiety is a temporary reaction to an event while a trait anxiety refers to a personality trait that predisposes an individual to experience anxiety (Spielberger et al., 1971). Within the context of climate change, individuals prone to trait anxiety can experience anxiety about climate events without ever having experienced one because they are innately predisposed to anxiety. Definitions of ecoanxiety across different disciplines (e.g. psychological sciences, anxiety studies) share characteristics such as ecoanxiety occurring in response to climate events featuring unpleasant feelings such as helplessness, depression, fear, and worry for one's future and that of future generations because of climate change.

Similar to ecoanxiety, researchers also describe a phenomenon called environmental or ecological grief (Cunsolo & Ellis, 2018; Kevorkian, 2004). Environmental grief refers to the feelings and emotions experienced in response to experienced or anticipated losses due to climate change, such as the loss of species, ecosystems, and meaningful landscapes or landmarks (Cunsolo & Ellis, 2018). Environmental grief is influenced by the psychosocial and cultural importance of these species, ecosystems, and landscapes and has the potential to effect both individuals and communities. Baudon and Jachens (2021) cited a quote from marine biologist Jaramillo (2019, as cited in Baudon & Jachens, 2021) who highlights the communal nature of environmental grief and describes the complexity of environmental grief: "I don't know where to go with my grief. Grief is a communal issue when it comes to climate change. Why would anyone grieve in solitude?" (p. 9). This shared aspect of environmental grief underscores its impact not only on individual well-being but also community mental health.

Children, young people, and Indigenous groups are the most vulnerable to experiencing ecoanxiety (Coffey et al., 2021). Research indicates that late childhood and early adolescence are key developmental periods for developing an interest in global environmental issues due to children at these developmental stages having the ability for abstract thinking and naturally begin to show an interest in broader societal issues (Evenshaug & Hallen, 2001, in Holden, 2007; Ojala, 2012). However, knowledge about global challenges like climate change may trigger feelings of anxiety, helplessness, and hopelessness (Ojala, 2012). This is particularly important for children who grow up around discussions of climate change and potential future consequences and, therefore, may have beliefs that climate change will cause the world to end in their lifetime (Albert et al., 2010, in Martin et al., 2022; Ojala, 2012). Martin et al. (2022)

reviewed 33 articles on the impact of climate change on school-aged children's (aged 3–19 years) mental well-being and negative emotions and found that negative emotions and poor mental well-being due to climate change were common. For example, 85% of Austrian and 89% of Australian adolescents reported that climate change was something to worry about (Harker-Schuch et al., 2021). Other studies on the impacts of climate change on children and their mental health indicated that as many as 2 in 3 children experienced at least moderate stress or anxiety about climate change, while 2 in 3 children at an outpatient psychiatric clinic reported symptoms of anxiety due to climate change (Baker et al., 2021; Kuang & Root, 2019, as cited in Martin et al., 2022). Additionally, the stress or anxiety children and young people felt about climate change was higher in adolescents approximately 15–18 years (Baker et al., 2021; Kuang & Root, 2019, as cited in Martin et al., 2022).

While there is a growing body of literature on children and young people in climate change research, Indigenous groups remain underrepresented. Unlike children and young people, research on the mental health impacts of climate change on Indigenous people is scarce. Yet, Indigenous groups experience amplified psychological impacts from climate change due to factors such as inhabiting ecologically sensitive areas and having their livelihoods closely tied to the environment (Middleton et al., 2020). Middleton et al. (2020) reviewed 50 articles published between 2007 and 2018 on the impacts of Indigenous mental health due to meteorological, seasonal, and climatic variables. The studies reviewed highlighted both acute and vicarious outcomes (Middleton et al., 2020).

For example, (Middleton et al., 2020) noted that Inuit women were experiencing fear and anxiety for friends and family members who were out during stormy conditions. The Inuit are Indigenous people of the Arctic and one of three recognised Indigenous Peoples in Canada where they are comprised of four regions: Inuvialuit Settlement Region, Nunatsiavut, Nunavik, and Nunavut. For many Inuit, wildlife harvesting like hunting and fishing make up a large part of their diet and economy. However, as the weather is more unpredictable and goes through changes from hot to cold more frequently due to climate change, the sea ice where Inuit hunters commonly hunt thaws and re-freezes more frequently. This leads to overall thinner and weaker ice and creates unsafe conditions for hunting (Dowsley et al., 2011). Additionally, due to rising global temperatures the ice breaks much earlier in the year such as mid-June versus the end of July—which significantly impacts the timeframe for the Inuit to hunt and gather resources. These conditions have led Inuit women to report greater levels of fear and anxiety for members of their tribes going out and encountering these dangerous conditions (Dowsley et al., 2011).

The indirect or vicarious impacts of climate change can be described as intense emotional responses like helplessness and fear known as ecoanxiety because of witnessing global environmental events. Children, adolescents, and Indigenous groups are particularly vulnerable to ecoanxiety with studies showing high levels of climate-related anxiety among these populations.

Psychosocial Impacts

The psychosocial impact refers to large-scale social and community effects that occur due to climate change such as heat-related violence, conflict over resources, environmental displacement, and post-disaster adjustment. Some of these impacts may occur directly due to climate change (e.g. heat-related violence; Clayton et al., 2021). Heat-related violence refers to the increase in aggressive behaviour that is observed during periods of increased temperature (Miles-Novelo & Anderson, 2019). This results as periods of heat increase stress in individuals which, in turn, makes individuals more irritable and more likely to react aggressively to provocation.

Children are increasingly vulnerable and oftentimes the targets of violence due to economic stress because of climate change (Cerna-Turoff et al., 2021). A study by Biswas et al. (2010) into unintentional injuries in children during flooding in rural Bangladesh highlighted a response by one child's father who admitted to slapping his child in response to an innocent request for a treat the family could not afford. The authors also noted that mothers were most often violent toward their children and were almost four times more likely to be violent if they did not earn an income outside of the household.

While some psychosocial impacts occur directly because of climate change, many are likely to be indirect effects such as challenges like increased stress, strained relationships, and grief and loss that arise due to displacement (Clayton et al., 2021). Environmental displacement refers to the forced, and often sudden, migration of individuals because of sudden or gradual climate events or human-induced environmental degradation or disasters (Kälin & Weerasinghe, 2017). Displacement often leaves individuals with limited social networks, living in overcrowded spaces, and having to compete for resources. Small Island Developing States (SIDS) are among some of the most prone to environmental effects (Plewa & Koch Giese, 2017). The Bahamas, an island nation in the Caribbean and one of the SIDS, is the third most affected country in the world by climate-related events (Eckstein et al., 2021). In 2019, the Bahamas were affected by hurricane Dorian, a category 5 hurricane and the most powerful hurricane ever to hit the island state. Dorian's impact caused the deaths of at least 70 individuals, an estimated US$ 3.4 billion in damages, and destroyed approximately 45% of homes on some islands (Eckstein et al., 2021; Marazita, 2020).

Hurricane Dorian also forced almost 10,000 people in the Bahamas to be displaced to locations within the islands of the Bahamas and externally to the United States and Canada (Marazita, 2020). These displacements did not only have geological challenges, but also raised issues of xenophobia, homophobia, and transphobia as the Haitian diaspora and members of sexual minority groups were among those displaced. There are an estimated 80,000 Haitian migrants in the Bahamas who often relocate for better employment opportunities and inhabit informal settlements. After the hurricane, more than 100 displaced Haitians were deported which led many of the remaining diaspora into hiding (International Organisation for Migration, 2019, as cited in Marazita, 2020). Accurate statistics on members of sexual minority groups who were displaced are difficult because there are no legal protections for this group in the Bahamas and they face threats of violence daily (Marazita, 2020). For example, the United States Human Rights Report on the Bahamas noted that sexual minority people in the Bahamas often faced social stigma and discrimination such as housing and employment-based discrimination (United States Department of State, n.d.).

The psychosocial impacts of climate change include heat-related violence, resource conflicts, and environmental displacement. Rising temperatures increase stress which can further increase aggression and aggressive behaviours. Economic challenges from climate change, like reduced crop yields, also increase stress and can lead to conflict over resources or violence against children, especially in lower income and struggling families. Environmental displacement forces individuals to move and can create or exacerbate social issues like xenophobia and homophobia as members of vulnerable groups potentially find themselves in new places with limited social networks and resources. These examples highlight some of the broad, often indirect, psychosocial impacts of climate change on communities.

Interventions for Climate-Related Psychosocial Challenges

Within the past decade, climate psychologists and other members of interdisciplinary teams have begun designing and implementing interventions to address the psychosocial impacts of anthropogenic climate change. Some examples of these interventions include counselling that can include therapy-based, community-based, or psychoeducation on climate literacy (Charlson et al., 2021; Monroe et al., 2019; Piggot-McKeller et al., 2019). Counselling or therapy-based interventions are used in cases of acute or vicarious impacts that manifest as emotional and mental health distress (Charlson et al., 2021). Similarly, community-based interventions and support networks are often used with groups of individuals who have already experienced the effects of climate change to boost their resilience against climate-related distress (Piggot-McKeller et al., 2019). Climate literacy interventions function similarly to community-based interventions by aiming to empower individuals and communities and boost their resilience toward climate-related distress but are often preventative instead of reactive (Monroe et al., 2019).

Counselling and Therapy-Based Interventions

Many counselling and therapy-based interventions are used to address acute or vicarious impacts of climate change such as PTSD, ecoanxiety, or depression (Charlson et al., 2021). For example, Baudon and Jachens (2021) reviewed 34 articles published between 2005 and 2019 on the treatment of ecoanxiety and noted recurring themes of building resilience, promoting climate action, and promoting a connection with nature. Interventions based on building resilience were done using a cognitive approach where the focus was on identifying and reframing individuals' thoughts, beliefs, and attitudes related to ecoanxiety. Cognitive interventions such as cognitive reappraisal or changing the way individuals think about the threat of climate change were particularly helpful with individuals who held black-and-white thoughts where ideas existed in two opposite extremes and in those prone to catastrophising about climate change. Building resilience was also facilitated through meaning-focused and existential interventions that helped individuals explore the relationship between their identity, values, and place in the world and ecoanxiety (Baudon & Jachens, 2021; Panno et al., 2015). Finally, resilience was also developed through emotion-focused interventions that helped individuals work through emotions like grief (Baudon & Jachens, 2021). Almost half of the emotion-based interventions reviewed, such as emotional expression or the acknowledgement and expression of feelings (Macy & Johnstone, 2012; Moser, 2012), focused on grief work and helping individuals experience mourning and loss without reliance on techniques like splitting, denial, and dissociation.

Resilience was also developed through nature-based interventions like climate action or encouraging individuals to reconnect with nature. Nature-based interventions to promote well-being, sometimes called 'green therapy,' have become increasingly popular, especially in urban spaces (Gabrielsen & Harper, 2017; Van den Berg, 2017). Climate action interventions were designed to support individuals in their efforts to connect, develop, and enact plans to reduce their carbon footprint (e.g. greenhouse gas emissions) and defend ecological values like energy efficiency and reducing consumption of meat and dairy. Interventions that focused on the individual's connection with nature appeared to function similarly to existential interventions by aiming to encourage individuals to connect with nature as a space of reflection and inspiration (Baudon & Jachens, 2021).

Community-Based Interventions and Support Networks

Community-based interventions and support networks aim to address the causes of vulnerabilities of communities and reduce the impacts of climate change (Piggot-McKeller et al., 2019). There is a strong emphasis on participation in these interventions and, as such, the members of these communities also become more empowered and resilient. Baudon and Jachens (2021) discussed community-based interventions, which encouraged groups to act against climate change or encouraged social connectedness and emotional support. Climate action groups usually function on multiple levels by providing a safe space for individuals to share similar experiences, beliefs, and attitudes which, in turn, reinforce and motivate action (Baudon & Jachens, 2021).

Community interventions based on social connectedness and emotional support were aimed at supporting individuals through the emotional process, like emotion-based counselling and therapy. Baudon and Jachens (2021) highlighted the need for these spaces, particularly in relation to environmental grief. Emotional support groups can function as safe spaces for the expression and containment for their distress about the future of the world, as well as having those feelings validated (Aston, 2022). Aston (2022) conducted a grounded theory investigation into the experiences of individuals who self-identified as living with ecoanxiety for at least two years. These individuals attended an ecopsychology support group for climate change for an 18-month period prior to the study. Aston identified four key themes within the data. The first theme was learning to be with difficult emotions that accompany an awareness of the climate crisis. Here, participants developed an emotional "toolkit" that allowed them to navigate and release the emotions they experienced because of ecoanxiety. Examples of strategies included attending support groups, receiving coaching from people who have similar experiences with ecoanxiety, and controlling their exposure to news and information about climate change. The second theme was building connections with people and nature toward resilience. This theme revolved around participants' enjoyment of finding community and bonding with like-minded individuals with shared values and beliefs. Many participants believed ecoanxiety was an isolating experience that could not be worked through alone and the climate crisis was a result of neglecting the fundamental human need for connection and instead relying on unsustainable capitalist lifestyles.

The third theme, Aston (2022), highlighted the theme of reconstructing and recalibrating participants' lives to their values and connections. A core belief within the sample was that social collapse because of climate change was imminent but there was no indication of when social collapse would happen. This knowledge influenced some participants' life decisions such as losing the desire to travel after retirement because of the impact of travel on the environment. Participants were forced to weigh the pros and cons of their needs against the needs of the environment which results in difficult decisions and sacrifices that is ultimately felt more congruent with their values. This led into the final theme: learning to welcome their grief. The author noted that participants described a concept called emotional tolerance, or the ability to recognise ecoanxiety as a valid response to something and messages that were important to listen to (Aston, 2022). Emotional tolerance allowed participants to cope with their anxiety because it allowed them to see ecoanxiety because of their love for life and vulnerable populations and their desire to reconnect with others. Acceptance was also a major component as participants came to terms with inspiring others to work against the climate crisis and they accept that as enough. Overall, Aston (2022) concluded that participants valued the group activity for making them feel heard and understood over a reduction in affect because feeling heard prompted them to grow and develop clarity on the actions they needed to take.

Climate Literacy-Based Interventions

Climate literacy-based interventions function similarly to community-based interventions by aiming to empower individuals and communities and boost their resilience toward climate-related distress. Other common objectives include improving knowledge about climate science, change, or solutions, or encouraging specific behaviours (Monroe et al., 2019). While these interventions increase one's knowledge about climate change and climate events, they also foster agency and preparedness, making individuals more equipped to handle the psychological impact of climate change (Monroe et al., 2019). Moreover, climate literacy-based interventions often make climate change personally relevant and meaningful to participants ensuring that participants can recognise the direct impacts on their lives and communities. Therefore, these interventions help build more informed and resilient individuals and communities who are better prepared for climate-related challenges (Monroe et al., 2019).

A review that evaluated climate change education interventions. More than half of these interventions were classroom-based as it took place in primary or secondary schools and tertiary education ($n = 28$) or universities ($n = 11$). Other locations included auditoriums and botanical gardens. These interventions were more effective if they made climate change information personally relevant and meaningful to learners and used activities that fostered engagement (Monroe et al, 2019). Climate change is often thought of as an ecological phenomenon with purely environmental consequences, and therefore, by making the individuals aware of the effects which leads to discussions on the impact of climate change in their own lives and communities, climate professionals can raise awareness and make individuals more likely to participate in climate action (Houghton et al., 2023).

Monroe et al. (2019) also expanded on some of the common teaching strategies used in these interventions and strategies to foster engagement. These were to have deliberate discussions, interact with scientists and the scientific process, address misconceptions about climate change, and design or implement school or community projects to address an aspect of climate change. Deliberate discussions helped learners to think more deeply, reflect, compare perceptions, and understand different opinions, while engaging with scientists and the scientific process helped to build engagement and have learners understand the processes involved in climate science. Some of these techniques were used simultaneously, as in one example a group of German students were able to address misconceptions about climate change and a hole in the ozone layer by simultaneously engaging in lab experiments that illustrated the greenhouse effect of carbon dioxide (Niebert & Gropengiesser, 2013).

Summary and Conclusion

Climate change is a natural process that takes place over hundreds of thousands of years. However, since the mid-eighteenth century, climate change has been developing much more rapidly due to human activity. Contrary to popular belief, the effects of climate change are widespread and have the potential to impact not only the environment but also human health and well-being and socioeconomic stability. In the past decade, there have been severe examples of the effects of climate change impacting communities' and wider countries' ecology, biodiversity, and agriculture industries, as well as impacts on human health through high mortality rates, pollution, infectious disease, and psychological distress.

The psychosocial impacts of climate change are felt directly by individuals and groups as evidenced by increased rates of PTSD, major depression, substance abuse, and self-harm and suicide (Cianconi et al., 2020). These impacts can also indirectly affect individuals as they observe the direct impacts on others and begin to internalise these events. This leads to what researchers have referred to as climate anxiety or ecoanxiety (Doherty & Clayton, 2011). These vicarious impacts are also known to disproportionately effect vulnerable groups like children and Indigenous communities. Other psychosocial impacts include increased rates of violence and conflict over resources, displacement, and challenges as individuals attempt to adjust to new environments following climate events.

As climate change has been developing over the past decade, so too have climate scientists and psychologists been responding to the needs of populations. Climate psychologists have been developing interventions for climate-related challenges and have been employing these on both individual and community levels (Charlson et al., 2021; Piggot-McKeller et al., 2019). Counselling and therapy-based interventions are often used to address acute and vicarious impacts, while community-based interventions have grown to specialise in working with individuals who suffer from ecoanxiety and climate-related-grief (Charlson et al., 2021; Piggot-McKeller et al., 2019). Climate literacy-based interventions are rooted in experiential learning and emphasise learner engagement and highlighting the personal relevance of climate change information (Monroe et al., 2019).

This chapter highlighted some recent literature on the psychosocial impacts of climate change. While research in this field has seen a rise in frequency, there is a substantial focus on experiences within the Global North, notably in countries like the United States, United Kingdom, Canada, and Australia. Furthermore, there is a glaring gap of literature from the most vulnerable groups such as Indigenous populations, individuals in low socioeconomic groups, and people of colour. Particularly noticeable was the scarcity of studies evaluating climate change interventions within these groups. These groups continue to experience disproportionate risks from the impacts of climate change. Therefore, there is a pressing need for further research aimed at developing targeted interventions tailored to the unique needs of these groups to address this critical gap in the literature.

References

American Psychiatric Association. (2013). *Diagnostic and statistical manual of mental disorders* (5th ed.). https://doi.org/10.1176/appi.books.9780890425596

American Psychological Association. (2023). Post-traumatic stress disorder (PTSD). In *APA dictionary of psychology*. Retrieved June 26, 2024, from https://dictionary.apa.org/posttraumatic-stress-disorder

Aston, N. (2022). *The Experience of eco-anxiety of individuals who have been through an ecopsychology support group for facing the climate crisis: A grounded theory inquiry*. [Doctoral thesis]. https://doi.org/10.15123/UEL.8V4QZ

Baker, C., Clayton, S., & Bragg, E. (2021). Educating for resilience: Parent and teacher perceptions of children's emotional needs in response to climate change. *Environmental Education Research, 27*(5), 687–705. https://doi.org/10.1080/13504622.2020.1828288

Baudon, P., & Jachens, L. (2021). A scoping review of interventions for the treatment of eco-anxiety. *International Journal of Environmental Research and Public Health, 18*(18), 9636. https://doi.org/10.3390/ijerph18189636

Biswas, A., Rahman, A., Mashreky, S., Rahman, F., & Dalal, K. (2010). Unintentional injuries and parental violence against children during flood: A study in rural Bangladesh. *Rural and Remote Health, 10*(1), 1199. https://pubmed.ncbi.nlm.nih.gov/20337500/

Boonstra, R., Bodner, K., Bosson, C., Delehanty, B., Richardson, E. S., Lunn, N. J., Derocher, A. E., & Molnár, P. K. (2020). The stress of Arctic warming on polar bears. *Global Change Biology, 26*(8), 4197–4214. https://doi.org/10.1111/gcb.15142

Born, D. (2019). Bearing witness? Polar bears as icons for climate change communication in *national geographic. Environmental Communication, 13*(5), 649–663. https://doi.org/10.1080/17524032.2018.1435557

Boudreau, D., Carangel, A., Costa, H., Jaszewski, J., McDaniel, M., Ramroop, T., Sprout, E., Teng, S., & Turgeon, A. (2022). *Earth's changing climate.* National Geographic Education. https://education.nationalgeographic.org/resource/earths-changing-climate/

Bucci, L., Alaka, L., Hagen, A., Delgado, S., & Beven, J. (2023). *Hurricane Ian (AL092022).* National Hurricane Center. https://www.nhc.noaa.gov/data/tcr/AL092022_Ian.pdf

Burke, M., González, F., Baylis, P., Heft-Neal, S., Baysan, C., Basu, S., & Hsiang, S. (2018). Higher temperatures increase suicide rates in the United States and Mexico. *Nature Climate Change, 8*(8), 723–729. https://doi.org/10.1038/s41558-018-0222-x

Cahill, A. E., Aiello-Lammens, M. E., Fisher-Reid, M. C., Hua, X., Karanewsky, C. J., Yeong Ryu, H., Sbeglia, G. C., Spagnolo, F., Waldron, J. B., Warsi, O., & Wiens, J. J. (2013). How does climate change cause extinction? *Proceedings of the Royal Society B: Biological Sciences, 280*(1750), 20121890. https://doi.org/10.1098/rspb.2012.1890

Carroll, B., Morbey, H., Balogh, R., & Araoz, G. (2009). Flooded homes, broken bonds, the meaning of home, psychological processes and their impact on psychological health in a disaster. *Health and Place, 15*(2), 540–547. https://doi.org/10.1016/j.healthplace.2008.08.009

Cerna-Turoff, I., Fischer, H. T., Mansourian, H., & Mayhew, S. (2021). The pathways between natural disasters and violence against children: A systematic review. *BMC Public Health, 21*, 1–17. https://doi.org/10.1186/s12889-021-11252-3

Charlson, F., Ali, S., Benmarhnia, T., Pearl, M., Massazza, A., Augustinavicius, J., & Scott, J. G. (2021). Climate change and mental health: A scoping review.

International Journal of Environmental Research and Public Health, 18(9), 4486. https://doi.org/10.3390/ijerph18094486

Chen, L., & Liu, A. (2015). The incidence of posttraumatic stress disorder after floods: A meta-analysis. *Disaster Medicine and Public Health Preparedness, 9*(3), 329–333. https://doi.org/10.1017/dmp.2015.17

Cheng, S., Plouffe, R., Nanos, S. M., Qamar, M., Fisman, D. N., & Soucy, J.-P.R. (2021). The effect of average temperature on suicide rates in five urban California counties, 1999–2019: An ecological time series analysis. *BMC Public Health, 21*(1), 974. https://doi.org/10.1186/s12889-021-11001-6

Christidis, N., & Stott, P. A. (2022). The extremely wet May of 2021 in the United Kingdom. *Bulletin of the American Meteorological Society, 103*(12), E2912–E2916. https://doi.org/10.1175/BAMS-D-22-0108.1

Cianconi, P., Betrò, S., & Janiri, L. (2020). The impact of climate change on mental health: A systematic descriptive review. *Frontiers in Psychiatry, 11*, 74. https://doi.org/10.3389/fpsyt.2020.00074

Clayton, S., Manning, C. M., Speiser, M., & Hill, A. N. (2021). *Mental health and our changing climate: Impacts, inequities, responses*. American Psychological Association and ecoAmerica. https://www.apa.org/news/press/releases/mental-health-climate-change.pdf

Clayton, S., Manning, C. M., Krygsman, K., & Speiser, M. (2017). *Mental health and our changing climate: Impacts, implications, and guidance*. American Psychological Association and ecoAmerica. https://www.apa.org/news/press/releases/2017/03/mental-health-climate.pdf

Coffey, Y., Bhullar, N., Durkin, J., Islam, M. S., & Usher, K. (2021). Understanding eco-anxiety: A systematic scoping review of current literature and identified knowledge gaps. *The Journal of Climate Change and Health, 3*, 100047. https://doi.org/10.1016/j.joclim.2021.100047

Coura, J. R. (2014). The main sceneries of Chagas disease transmission. The vectors, blood and oral transmissions—A comprehensive review. *Memórias Do Instituto Oswaldo Cruz, 110*(3), 277–282. https://doi.org/10.1590/0074-0276140362

Cunsolo, A., & Ellis, N. R. (2018). Ecological grief as a mental health response to climate change related loss. *Nature Climate Change, 8*(4), 275–281. https://doi.org/10.1038/s41558-018-0092-2

Doherty, T. J., & Clayton, S. (2011). The psychological impacts of global climate change. *American Psychologist, 66*(4), 265–276. https://doi.org/10.1037/a0023141

Dowsley, M., Gearheard, S., Johnson, N., & Inksetter, J. (2011). Should we turn the tent? Inuit women and climate change. *Études/inuit/studies, 34*(1), 151–165. https://doi.org/10.7202/045409ar

Eckstein, D., Kunzel, V. & Schafer, L. (2021). *Global Climate Risk Index 2021*. GermanWatch. https://www.germanwatch.org/en/19777

Fontalba-Navas, A., Lucas-Borja, M. E., Gil-Aguilar, V., Arrebola, J. P., Pena-Andreu, J. M., & Perez, J. (2017). Incidence and risk factors for post-traumatic stress disorder in a population affected by a severe flood. *Public Health, 144*, 96–102. https://doi.org/10.1016/j.puhe.2016.12.015

Gabrielsen, L. E., & Harper, N. J. (2017). The role of wilderness therapy for adolescents in the face of global trends of urbanization and technification. *International Journal of Adolescence and Youth*, 1–13. https://doi.org/10.1080/02673843.2017.1406379

Gay, C., Estrada, F., Conde, C., Eakin, H., & Villers, L. (2006). Potential impacts of climate change on agriculture: A case of study of coffee production in Veracruz, Mexico. *Climatic Change, 79*(3–4), 259–288. https://doi.org/10.1007/s10584-006-9066-x

Golitaleb, M., Mazaheri, E., Bonyadi, M., & Sahebi, A. (2022). Prevalence of post-traumatic stress disorder after flood: A systematic review and meta-analysis. *Frontiers in Psychiatry, 13*, 890671. https://doi.org/10.3389/fpsyt.2022.890671

Harker-Schuch, I., Lade, S., Mills, F., & Colvin, R. (2021). Opinions of 12 to 13-year-olds in Austria and Australia on the concern, cause and imminence of climate change. *Ambio, 50*(3), 644–660. https://doi.org/10.1007/s13280-020-01356-2

Hoell, A., Quan, X.-W., Hoerling, M., Diaz, H. F., Fu, R., He, C., Lisonbee, J. R., Mankin, J. S., Seager, R., Sheffield, A., Simpson, I. R., & Wahl, E. R. (2022). Water year 2021 compound precipitation and temperature extremes in California and Nevada. *Bulletin of the American Meteorological Society, 103*(12), E2905–E2911. https://doi.org/10.1175/BAMS-D-22-0112.1

Holden, C. (2007). Young people's concerns. In D. Hicks & C. Holden (Eds.), *Teaching the global dimension: Key principles and effective practice* (pp. 31–42). Routledge.

Houghton, S., Garvey, J., Conor, L., Wilmsen, B., Dehm, J., Gamble, R., Habib, B., Holmes, K., Millner, J., & Strickland, K. M. (2023). Towards an interdisciplinary agenda for teaching in the climate crisis: Reflections from the humanities and social sciences. *Environmental Education Research*, 1–13. https://doi.org/10.1080/13504622.2023.2273791

Intergovernmental Panel on Climate Change. (2021). *Climate change widespread, rapid, and intensifying—IPCC*. https://www.ipcc.ch/2021/08/09/ar6-wg1-20210809-pr/

Intergovernmental Panel on Climate Change (IPCC). (2023). Changing state of the climate system. In *Climate Change 2021—The Physical Science Basis:*

Working Group I Contribution to the Sixth Assessment Report of the Intergovernmental Panel on Climate Change (pp. 287–422). chapter, Cambridge: Cambridge University Press.

Kalin, W. & Weerasinghe, S. (2017). *Environmental migrants and global governance: Facts, policies and practices.* International Organization for Migration. https://publications.iom.int/es/system/files/pdf/migration_research_leaders_syndicate.pdf

Kevorkian, K. A. (2004). *Environmental grief: Hope and healing.* Union Institute & University.

Knutson, T. R., Chung, M. V., Vecchi, G., Sun, J., Hsieh, T.-L., & Smith, A. J. P. (2021). Climate change is probably increasing the intensity of tropical cyclones. In C. Le Quere, P. Liss, & P. Forster (Eds.), *Critical issues in climate change science.* https://doi.org/10.5281/ZENODO.4570334

Koenen, K. C., Ratanatharathorn, A., Ng, L., McLaughlin, K. A., Bromet, E. J., Stein, D. J., Karam, E. G., Meron Ruscio, A., Benjet, C., Scott, K., Atwoli, L., Petukhova, M., Lim, C. C. W., Aguilar-Gaxiola, S., Al-Hamzawi, A., Alonso, J., Bunting, B., Ciutan, M., De Girolamo, G., Kessler, R. C., et al. (2017). Posttraumatic stress disorder in the world mental health surveys. *Psychological Medicine, 47*(13), 2260–2274. https://doi.org/10.1017/S0033291717000708

Lertzman, R. A. (2015). *Environmental melancholia: Psychoanalytic dimensions of engagement.* Routeledge.

Lorentzou, C., Kouvarakis, G., Kozyrakis, G. V., Kampanis, N. A., Trahanatzi, I., Fraidakis, O., Tzanakis, N., Kanakidou, M., Agouridakis, P., & Notas, G. (2019). Extreme desert dust storms and COPD morbidity on the island of Crete. *International Journal of Chronic Obstructive Pulmonary Disease, 14*, 1763–1768. https://doi.org/10.2147/COPD.S208108

Macy, J., & Johnstone, C. (2012). *Active hope: How to face the mess we're in without going crazy.* New World Library.

Malhi, G. S., Kaur, M., & Kaushik, P. (2021). Impact of climate change on agriculture and its mitigation strategies: A review. *Sustainability, 13*(3), 1318. https://doi.org/10.3390/su13031318

Marazita, J. (2020). *Displacement in paradise: Hurricane Dorian slams the Bahamas.* Internal Displacement Monitoring Centre. https://www.internal-displacement.org/publications/displacement-in-paradise-hurricane-dorian-slams-the-bahamas/

Martin, G., Reilly, K., Everitt, H., & Gilliland, J. A. (2022). Review: The impact of climate change awareness on children's mental well-being and negative emotions—a scoping review. *Child and Adolescent Mental Health, 27*(1), 59–72. https://doi.org/10.1111/camh.12525

Materia, C. J. (2016). *Climate state anxiety and connectedness to nature in rural Tasmania.* [Unpublished doctoral thesis]. University of Tasmania.

Middleton, J., Cunsolo, A., Jones-Bitton, A., Wright, C. J., & Harper, S. L. (2020). Indigenous mental health in a changing climate: A systematic scoping review of the global literature. *Environmental Research Letters, 15*(5), 053001. https://doi.org/10.1088/1748-9326/ab68a9

Miles-Novelo, A., & Anderson, C. A. (2019). Climate change and psychology: Effects of rapid global warming on violence and aggression. *Current Climate Change Reports, 5*(1), 36–46. https://doi.org/10.1007/s40641-019-00121-2

Mollica, N. R., Guo, W., Cohen, A. L., Huang, K.-F., Foster, G. L., Donald, H. K., & Solow, A. R. (2018). Ocean acidification affects coral growth by reducing skeletal density. *Proceedings of the National Academy of Sciences, 115*(8), 1754–1759. https://doi.org/10.1073/pnas.1712806115

Monroe, M. C., Plate, R. R., Oxarart, A., Bowers, A., & Chaves, W. A. (2019). Identifying effective climate change education strategies: A systematic review of the research. *Environmental Education Research, 25*(6), 791–812. https://doi.org/10.1080/13504622.2017.1360842

Moser, S. (2012). *Getting real about it: Meeitng the psychological and social demands of a world in stress.* https://climateaccess.org/resource/getting-real-about-it-meeting-psychological-and-social-demands-world-stress/

National Aeronautics and Space Administration. (n.d.). *What is the greenhouse effect?* https://science.nasa.gov/climate-change/faq/what-is-the-greenhouse-effect/

National Oceanic and Atmospheric Administration. (2019). *Carbon cycle.* Retrieved October 15, 2023, from https://www.noaa.gov/education/resource-collections/climate/carbon-cycle

National Oceanic and Atmospheric Administration Fisheries. (2023). *Seals, sea lions, and climate change: Shifting prey and habitat impacts.* Retrieved October 15, 2023, from https://www.fisheries.noaa.gov/national/climate/seals-sea-lions-and-climate-change-shifting-prey-and-habitat-impacts

Niebert, K., & Gropengiesser, H. (2013). Understanding and communicating climate change in metaphors. *Environmental Education Research, 19*(3), 282–302. https://doi.org/10.1080/13504622.2012.690855

Ojala, M. (2012). Regulating worry, promoting hope: How do children, adolescents, and young adults cope with climate change? *International Journal of Environmental and Science Education, 7*(4), 537–561. https://eric.ed.gov/?id=EJ997146

Panno, A., Carrus, G., Maricchiolo, F., & Mannetti, L. (2015). Cognitive reappraisal and pro-environmental behavior: The role of global climate change perception. *European Journal of Social Psychology, 45*(7), 858–867. https://doi.org/10.1002/ejsp.2162

Pham, Y., Reardon-Smith, K., Mushtaq, S., & Cockfield, G. (2019). The impact of climate change and variability on coffee production: A systematic review.

Climatic Change, 156(4), 609–630. https://doi.org/10.1007/s10584-019-02538-y

Piggott-McKellar, A. E., McNamara, K. E., Nunn, P. D., & Watson, J. E. M. (2019). What are the barriers to successful community-based climate change adaptation? A review of grey literature. *Local Environment, 24*(4), 374–390. https://doi.org/10.1080/13549839.2019.1580688

Pihkala, P. (2020). Anxiety and the ecological crisis: An analysis of eco-anxiety and climate anxiety. *Sustainability, 12*(19), 7836. https://doi.org/10.3390/su12197836

Plewa, M. P. & Koch Giese, I. (2017). *Disaster-induced displacement: The case of small island developing states in the Pacific and the Caribbean* [Unpublished research project]. Lund University.

Pörtner, H.-O., Roberts, D. C., Adams, H., Adelekan, I., Adler, C., Adrian, R., Aldunce, P., Ali, E., Ara Begum, R., BednarFriedl, B., Bezner Kerr, R., Biesbroek, R., Birkmann, J., Bowen, K., Caretta, M. A., Carnicer, J., Castellanos, E., Cheong, T. S., Chow, W., Zaiton Ibrahim, Z., et al. (2022). *Technical Summary.* In H.-O. Pörtner, D. C. Roberts, M. Tignor, E. S. Poloczanska, K. Mintenbeck, A. Alegría, M. Craig, S. Langsdorf, S. Löschke, V. Möller, A. Okem, B. Rama (Eds.), *Climate Change 2022: Impacts, Adaptation and Vulnerability. Contribution of Working Group II to the Sixth Assessment Report of the Intergovernmental Panel on Climate Change.* Cambridge University Press: Cambridge and New York: NY. https://doi.org/10.1017/9781009325844.002

Santos, V. R. C. D., Meis, J. D., Savino, W., Andrade, J. A. A., Vieira, J. R. D. S., Coura, J. R., & Junqueira, A. C. V. (2018). Acute Chagas disease in the state of Pará, Amazon region: Is it increasing? *Memórias Do Instituto Oswaldo Cruz, 113*(5). https://doi.org/10.1590/0074-02760170298

Schweitzer, M. D., Calzadilla, A. S., Salamo, O., Sharifi, A., Kumar, N., Holt, G., Campos, M., & Mirsaeidi, M. (2018). Lung health in era of climate change and dust storms. *Environmental Research, 163*, 36–42. https://doi.org/10.1016/j.envres.2018.02.001

Sousa Júnior, A. D. S., Palácios, V. R. D. C. M., Miranda, C. D. S., Costa, R. J. F. D., Catete, C. P., Chagasteles, E. J., Pereira, A. L. R. R., & Gonçalves, N. V. (2017). Análise espaço temporal da doença de Chagas e seus fatores de risco ambientais e demográficos no município de Barcarena, Pará, Brasil. *Revista Brasileira De Epidemiologia, 20*(4), 742–755. https://doi.org/10.1590/1980-5497201700040015

Spielberger, C. D., Gonzalez-Reigosa, F., Martinez-Urrutia, A., Natalicio, L. F. S., & Natalicio, D. S. (1971). Development of the Spanish Edition of the state-trait anxiety inventory. *Interamerican Journal of Psychology, 5*(3–4), 145–158. https://doi.org/10.30849/rip/ijp.v5i3%20&%204.620

United Nations. (n.d.a). *What is climate change*. https://www.un.org/en/climatechange/what-is-climate-change

United Nations. (n.d.b). *Causes and effects of climate change*. https://www.un.org/en/climatechange/science/causes-effects-climate-change

United States Department of State. (n.d.). *2022 country reports on human rights practices: The Bahamas*. https://www.state.gov/reports/2022-country-reports-on-human-rights-practices/the-bahamas/

Van Den Berg, A. E. (2017). From green space to green prescriptions: Challenges and opportunities for research and practice. *Frontiers in Psychology, 8*, 268. https://doi.org/10.3389/fpsyg.2017.00268

Williams, M. N., Hill, S. R., & Spicer, J. (2016). Do hotter temperatures increase the incidence of self-harm hospitalisations? *Psychology, Health and Medicine, 21*(2), 226–235. https://doi.org/10.1080/13548506.2015.1028945

World Health Organisation. (2015). *Biodiversity and health*. https://www.who.int/news-room/fact-sheets/detail/biodiversity-and-health

World Health Organisation. (2022). *Mental health and climate change: Policy brief*. https://www.who.int/publications/i/item/9789240045125

World Health Organisation. (2023). *Climate change*. https://www.who.int/news-room/fact-sheets/detail/climate-change-and-health

World Health Organisation. (2024). *Chagas disease (also known as American trypanosomiasis)*. https://www.who.int/news-room/fact-sheets/detail/chagas-disease-(american-trypanosomiasis)

CHAPTER 4

The Climate-Resilient Elderly of Barbados

Melony Shoye

Introduction

Climate variability makes predictions of the effects of hydro-meteorological events on island-states challenging. The effects of climate variability in the Caribbean can be felt with increasingly reduced annual precipitation and months of sudden and increasingly destructive storm events. Climate variability poses a particular risk to the exposed coastline of the small island developing state (SIDS) of Barbados. The sources of Barbados' floodwaters can be found in two main types of hydro-meteorological events. That is, rainfall and coastal storm surges. Both events generally occur during the locally known "wet season" of the year, between the months of June to November. During heavy rainfall, surface water runs unidirectionally from inland towards the sea. It flows through the island's multi-veined, extensive gully systems (Cumberbatch et al. 2020; DEM 2020; Donovan & Harper, 2005; Evanson, 2014). These storm events contribute to flooding (Biguenet et al., 2023; Kirton, 2013; Sullivan, 2017; Thomas et al., 2020).

M. Shoye (✉)
The University of the West Indies, Bridgetown, Barbados
e-mail: melonyshoye@gmail.com

D. D. Joseph et al. (eds.), *Climate Change*,
https://doi.org/10.1007/978-3-031-83037-2_4

Lighter but consistent rainfall tends also to saturate the water table, causing the build-up of surface run-offs towards the sea, through natural waterways. Physical development on the island affects the easy flow of water through the gullies, as the building of roadways and other infrastructure, or the dumping of garbage, can narrow or restrict the water passages. Flooding occurs in these circumstances. In storm surge events, water flows in the opposite direction from the sea surging towards the island's coastal towns. This is where flooding also occurs. The flood plains around the island in this case are identified (not exhaustively) as the Speightstown, Holetown, Bridgetown, Oistins and Belleplaine areas (Cumberbatch et al. 2020; DEM 2020; Donovan & Harper, 2005; Evanson, 2014). These flood events pose a particular risk to the exposed, climate change-sensitive environment of Barbados, particularly for the elderly. It is a hazard that further exposes the already-vulnerable demographic of the elderly living in Barbados to dire health and socio-economic risks.

Global Overview

Globally, there is a concern over the elderly's resilience response to flooding events as they are living longer yet their quality of life is generally expected to be marked by physical and mental decline. The Pan American Health Organization (PAHO, 2023) advocates for greater inclusion of the elderly in disaster management, emphasizing their valuable contributions to community resilience. Both males and females 60 years and older experience increased frailty, reduced mobility, incontinence and balance as well as susceptibility to infectious diseases. They are the least able to cope in the event of disasters such as flooding, not only because of the above-mentioned challenges, but because they are also easy victims of violence and fraud in the aftermath of disasters (Beard & Bloom, 2015; ben Hassen et al., 2019; Berry et al., 2009, Johnson et al., 2014). This chapter examines the lived experience of the Barbadian elderly in two, locally known epicentres for flooding events on the island: the rural community of Belleplaine, St. Andrew and a community within the urban corridor, Holetown, St. James.

Barbados

Barbados is a small island nation in the eastern Caribbean with a population of approximately 287,000 people. The tiny island is 430 square kilometres (166 square miles) in size and most of the population, about 92%, is of African descent. English is the official language with a local Bajan dialect, which blends English with African linguistic influences. Twenty per cent of the population is under the age of 15, and one in five persons is expected to be over the age of sixty years in Barbados by 2025 (Brown & Murphy, 2018; Downes et al., 2015). By 2050, this percentage of persons over the age of 60 years is expected to be 35.4% of the population (Devonish & Nurse, 2023). In Barbados, the life expectancy is 62.3 years (Devonish & Nurse, 2023), with a high mortality rate due to chronic non-communicable diseases including diabetic-related illnesses such as amputation and diabetes mellitus, hypertension, coronary disease, cancer and stroke (Springer, 2012). In 2015, the count was that "one in two of those aged 65 years or older" was diabetic (Devonish & Nurse, 2023; Hambleton et al., 2005; ILO 2022; Springer, 2012). There is also a rising presence of Alzheimer's disease among the elder population of Barbados. By 2050, it is expected that 8000 elderly persons will be living with dementia (Devonish & Nurse, 2023). In flood events, elderly persons with disabilities and chronic diseases are among the most vulnerable. The elderly's resilience to adverse climate events has rarely been examined. For the elderly, resilience is defined by the ability to prepare for and adapt to changing conditions and withstand and recover rapidly from disruptions. Resilience includes the ability to withstand and recover from disturbances such as an attack, an accident or naturally occurring threats or incidents. Currently, there is no known study that has examined the Barbadian elderly and their resilience to withstand and recover from flood events in a timely manner.

Effects of Flood on the Elderly

The effects of flood events on the elderly can vary based on several factors, including physical displacement due to the loss of their homes from raging surface waters, the risk of death or severe injury during the flood, and increased susceptibility to illness and injury in its aftermath. Elderly persons who are socially isolated and do not have sufficient financial resources to repair their homes or are able to purchase medication or transport themselves to medical services, if needed, are particularly vulnerable and less resilient after a flood event (Chavez-Alvarado and Diego-Gonzalez, 2016; Dhemba, 2012, Dhemba & Dhemba, 2015). It is important to understand the factors that influence and measure resiliency among the elderly to recover from the trauma of flood events and return to their previous level of functioning (Greene, 2002; Greene & Cohen, 2005; Greene & Greene, 2022), which is not widely explored in research literature.

This chapter examines elderly residents living in two communities in Barbados: Belleplaine, St. Andrew, which is located on the rural east coast, and Holetown, St. James, which is located on the urban corridor of the west coast of the island. The Draft Physical Development Plan (TCDPO, 2023) for Belleplaine, St. Andrew highlights its centrality as a National Park hub for sustainable nature tourism. It is an area covered in slopes of land of sandstone and other unstable soil features which flatten out to a populated plain. Belleplaine has a river system which operates as a natural stormwater catchment.

Holetown, St. James is described as even more flood-prone. Several gully catchments of the Folkestone Branch, Portvale, Seaview and Molyneux, with other gully outlets identified at Holetown Lagoon and north of the civic centre, empty into the built-up urban area creating such issues as the overwhelming of drainage systems during rainfall events (Limonta et al., 2015; TCPDO Draft, 2023).

In Barbados, there have been increasingly organized improvements in the approach to disaster management since the 1970s. It has evolved beyond the pre-1970s approach of "response and relief" to the people affected by hurricanes. The prevention, mitigation, preparedness, response, recovery and rehabilitation (PMPRRR) methodology is a comprehensive framework designed to manage the entire disaster lifecycle. It includes strategies for reducing the risk of disasters, preparing for potential impacts, responding effectively when disasters occur, and facilitating recovery and rehabilitation efforts in the aftermath. However, while this methodology provides a structured approach, it often employs a top-down, where decisions are made at higher levels with little input from specific communities and people. The unique needs of the elderly regarding their housing conditions, their needs for social support and their viability with regard to their finances and health are factors that affect an elderly person's potential for disaster preparedness and resilience.

Since 2007, the Barbados Department of Emergency Management has oversight for the prevention and mitigation of floods and other hazard events. The collaboration with other technical and social agencies, such as the Barbados Meteorological Services, the Coastal Zone Management Unit and the Barbados Welfare Department, provides access to rainfall data, flood inundation maps and social care responses, island wide. There is also collaboration with the larger, Caribbean-wide agency, the Caribbean Disaster Emergency Management Agency (CDEMA), in a regional response mechanism to disaster management and the United States National Hurricane Centre (NHC) which tracks tropical disturbances coming west across the Atlantic Ocean. There is, however, not a specific elder resilience focus made by these entities.

Social Work Interventions

Social work interventions for the elderly, in response to the hazard of flooding, are mainly government-led. Most Barbadian social workers are civil servants who perform several roles. They are educators to elderly clientele on the need to prepare for natural disasters. During post flood events, they perform damage assessments, visiting reported sites where persons receive storm damage to their homes. They are trauma counsellors, assisting the elderly with grief-related issues due to losses during a climate event. They are facilitators of access to food, furnishings and other forms of assistance in kind resources. They also liaise with other social service agencies for elderly persons who need housing or temporary rental accommodation (Devonish & Nurse, 2023; Springer, 2012). However, these services are not enough to address the needs of the elderly during a flood event as there are many gaps in local flood preparedness that do not acknowledge the experience of elderly Barbadians and their potential to contribute to the discussion. One aspect of this contribution can be captured using a phenomenology-based model that highlights the lived experiences of the elder person. A second aspect of this contribution relies on the social worker in the community to be the mediator between the elderly and the flood management technicians, while being an advocate for the relevance of the local knowledge which the elderly can share.

While disaster management in Barbados has evolved with more organized approaches and frameworks like PMPRRR, significant gaps remain in addressing the specific needs of elderly individuals in disaster preparedness and resilience.

Environmental Justice and Resilience Theory

To effectively address the needs of the elderly, this research employed the theoretical frameworks of environmental justice and resilience theory within the context of social work. Resiliency theory includes an understanding of adversity which can vary in duration and intensity, ranging from short- to long-term outcomes. Responses to adversity differ based on individual experiences, geographical context and definitions of "good outcomes." Van Breda (2018) emphasizes recognizing patterns in adversity responses. It also includes mediating processes which play a crucial role in resilience which is the balance between protective resources, such as family, community support and individual traits like optimism. Another aspect of resilience is the outcomes which can vary as it is measured on a spectrum rather than as binary success or failure. Ungar (2018) expands on this by highlighting the importance of environmental factors and protective processes, including social structures and services that help reduce risks and foster resilience. Environmental justice emphasizes fair treatment and meaningful involvement, ensuring equal access to decision-making on environmental issues and equal protection from environmental hazards. This concept, coined by the U.S. Environmental Protection Agency (Beltran et al., 2016; Joseph & Doon, 2022; Rambaree, 2020), advocates for equitable distribution, where all individuals experience equal risks and benefits from their environment and have equal access to resources to meet their needs.

Social workers play a key role in advocating for vulnerable groups, such as the elderly, by addressing these issues on both policy and practice levels through activism and ensuring justice in environmental concerns. Both frameworks offer personalized perspectives on how individuals and social groups interact with their environment, the stressors or hazards they encounter, and how these interactions shape their lived experiences. The intersection of these theories highlights the importance of recognizing the strengths of individuals and communities in their capacity to positively respond to challenges. Ungar's (2004) resilience theory and environmental justice emphasize on allowing minority social groups to culturally define their own experiences with climate-related events. These two theories offer a context of elder resilience which is characterized by factors such as religiosity or spirituality, and the support provided by family, community and organizations (Brockie & Miller, 2017; Goddard-Durant, 2019; Lacet, 2016; Silva et al., 2019).

Methods—Study Design

This paper's qualitative approach leans on the hermeneutic phenomenology of Pietkiewicz and Smith (2014) in describing an approach that "investigate[s] how individuals make sense of their experiences" by "focusing on how people perceive and talk about objects and events" (p 8). Hermeneutic phenomenology involves a detailed examination of an individual's description of their experience of a phenomenon and their understanding of resilience. This research concentrates in two communities in Barbados: Belleplaine, St. Andrew, which is located on the rural east coast, and Holetown, St. James, which is situated in the urban corridor of the west coast of the island. The research questions examined the lived experience of elder persons living in Belleplaine, St. Andrew and Holetown, St. James and how the elderly negotiated receiving assistance during the flood events.

Researcher Assumptions

As a scholar practitioner, I currently work as a social worker in the Welfare Department and have experience working with the elderly during a flood event. This experience alongside the findings of the literature review found the elderly will mainly live alone and/or have a very limited kinship or community support network. In some cases, the elderly may be too frail to maintain resilience on their own. Some elderly despite experiencing previous flood event will continue to reside in flood-prone areas because of their ties to the land/community.

Recruitment

Participants were recruited using purposive sampling of a non-probability sample and snowball sampling which is suitable for phenomenological research (Babbie, 2010). Persons were initially selected from the client list of the Barbados Welfare Department. From this starting point, further participants were selected using the snowballing method to saturate the study with all possible lived experiences of resilience. The point of saturation is reached when there are no further new statements/themes given by the participants on their experience. An invitation letter was sent to participants. The participants were contacted before to schedule the interviews. The potential participant and their caregivers, if any, were informed on the particulars of the study. These particulars include "its duration, its activities, location, how much time will be required and that the interview will be recorded." Two informed consent forms were given to the participant for them to sign after the interview was performed. Participants will keep one of these forms. Caregivers were given the opportunity to preview the form and raise any objections, if necessary. An interview protocol was followed with open-ended questions. Data collection involved conducting interviews, the taking of field notes and providing home visits (Groenewald, 2004) (Table 4.1).

Procedure

Semi-structured interviews were conducted with eight older residents, four living in Belleplaine, St. Andrew and four living in Holetown, St. James. Data collection occurred in 2021, during stay-at-home restrictions due to COVID-19. When possible, in-person interviews were conducted to examine the body language of the participant. Due to the restrictions, other interviews were done via telephone, as well as WhatsApp video conferencing. Each interview was 45 minutes in length and when needed a support person was present with the elderly person, if desired, especially to facilitate any social media video conferencing. The support person did not answer on behalf of the participant unless in rare cases to provide explanations where the researcher is unable to understand the response adequately. In this case, reference was made to the participant to ensure that his/her exact meaning is being conveyed. Participants signed a consent form, and the study was approved by the University of the West Indies Cave Hill ethics committee. All participants were asked about their understanding of flood events, their understanding of accessing support post a flood event and the time lapse between asking for assistance and receiving it.

Table 4.1 Participants by gender, age, health conditions, place of residence and length of time residing in the community

Gender	*Health conditions*	*Place of residence*	*Length of time living in the area*	*Date of birth/ age*
Female		Jamestown Park, St. James	40 years; resides with family; very physically active	68 years old
Female		Belleplaine, St. Andrew (Belleplaine housing area)	Moved into the area in 1969; moved into current home in 1998; matriarch of a household of young adults and children	1957
Female		Franklin Douglin Road in Belleplaine, St. Andrew	Always resided there but now living in a different house next-door to her daughter	1950
Female		Lives in Holetown # 2; moved from Trents, St. James to current home	55 years; resides with a female relative of similar age	67 years old
Female	Cancer survivor	Worked in Holetown for 50 years; Lives in a different community in St. James	Started working in Holetown at 15 years old at a large supermarket business and remained for 50 years. Resides with her husband who has dementia. In daily contact with friends and relatives	Started working in December, 1969/finished working in 2019/67 years old

(continued)

Table 4.1 (continued)

Gender	*Health conditions*	*Place of residence*	*Length of time living in the area*	*Date of birth/ age*
Female	Amputee	Trents, St. James	Lived in Discovery, St. James as a child; was relocated to Trents, St. James after a devastating flood. Later relocated to 1st Street, St. James where her house received flood damage and is now living back in Trents, St. James in her daughter's home	70 years old
Male		Belleplaine, St. Andrew	Lived in the area for 71 years. Relies on neighbours to run errands for him	77 years old
Male	Very poor eyesight	Belleplaine, St. Andrew	Resides alone. His son resides nearby and visits daily to give assistance. House has no electricity	84 years old

Analysis

All interviews were recorded and transcribed verbatim, deidentified and uploaded to NVIVO qualitative analysis software for data management and analysis. The data were analysed using a narrative analysis to identify key themes across the interviews. Codes were identified, and data extracts were grouped and collated within the individual codes. The subsequent generation of sub-themes and overarching themes is an iterative process that involves re-reading, coding and recoding until overarching themes are realized. The researcher maintained reflexivity by having debriefing sessions with an independent supervisor familiar with this area of research.

Results

True to Heidegger's (Dilby et al., 2020) argument, there is validity in personal definitions of a phenomenon within socially researched contexts. There were three themes surrounding the resilience of elderly participants in these communities.

1. They have a trust in God.
2. They know the geography of their communities.
3. They have an immediate, reliable social support network of relatives and friends as well as a reliance on the provision of the government regarding flood preparedness and recovery.

Theme I—They have a Trust in God

The following quotes showed that more than one reference was made to the theme entitled "Having a trust in God," indicating a shared understanding among those participants.

For the sixty-four-year-old woman living in Belleplaine, St. Andrew, her trust in God eliminated fear of flooding and its effects on her physical environment,

> I am not afraid. All I say the Almighty, I say Lord, You is my Provider, You will take care of me.

For a sixty-seven-year-old woman living in Holetown, St. James, she pointed to evidence of what God has done in the past, as the foundation on which she placed her present trust,

> ...when I get [the] storm here, we always pray to the Lord and He always save us.

Praying was an automatic, active necessity for another sixty-seven-year-old female who worked in Holetown for fifty years, especially when the area was threatened by hydro-meteorological events and there was no other consistent messaging informing her safety,

> That time [you don't have a] radio... [You] would hear [news] in between but when you look outside and... you see outside look dark, you tell yourself, 'oh'... If you see certain things you [start] praying, praying that...everything good and nothing [bad would] happen. [Yes, you] pray hard girl. [You] used to pray hard hard.

For a female, seventy-one-year-old resident of Belleplaine, she lived by a principle that was founded in the statement,

> All you gotta say, put God in front. And He will direct you[r] path". Her statement made reference to the King James Version of the Christian bible, in Proverbs 3: 5–6 which said, "Trust in the Lord with all your heart, And lean not on your own understanding; In all your ways acknowledge Him, And He shall direct your paths.

Lacet (2016) supported the findings that spirituality was an important cultural factor because it was recognized as an important part of at least half of the participants' lives.

Theme 2—They Know the Geography of their Communities

This sixty-four-year-old woman from Belleplaine, St. Andrew could trace the direction of the water course in rainfall event. She was able to identify areas where houses were safely situated.

> ...water...comes like from up in... St Joseph, comes down the river ... then ... water [comes] from Hillaby, to White Hill, and meets back there to Bruce Vale. Then it comes [all] the way down, then meet Coggins. Then it comes right down, meets then to East Coast, [and flows across] East 110 Coast and then come down.

Another resident, seventy-one years old, who lived in the sloped land above the Belleplaine main road, added a knowledge of the effect of topography on water flow,

> If the water flow and go...you see where this house here [gestures with her hand] It can't affect them [the residents in that area]... [that place is] something like a gully... so it can't affect them [because] them [are situated on] the hill... It would rush and go [a]long... down that way

> and [becomes] lost… inside them trees… It just like a gully… That water would rush through the gully [until it reaches] Long Pond.

One Holetown, St. James female resident, sixty-seven years old, could also identify one of the several watercourses making its way to her community,

> …the water used to come from… up in St. Joseph… and St. Andrew and come straight down here and full up all there [gestures towards a place in the community] and it come over the road. And… full up the river there so by the [police] station…

A second female participant who worked in Holetown, St. James for 50 years,

> The water from St. Thomas used to go down to Holetown… [Because] you know…when [you] come from St. Thomas Church, it would come down… over that hill and it would [go] down…to the Methodist Church… and then it would come from the Methodist Church and…go up to the Post Office. Some [of the water] used to go down to [where] Limegrove [the fancy shopping mall, is] now…and then it would go straight up to the Police Station and go in the Police Station, then it would… go cross like if it [going] through to Cherry Avenue [above the supermarket], and then it would come in to the Supermarket.

Third Theme—Having a Reliable Social Support Group of Family or Friends and on the Government's Flood Recovery Process

The participants spoke of the role to families and community persons pursuing their own recovery efforts. Both themes were often presented together in the interviewees' responses. The reliance on a household's collective effort usually would take precedent over a reliance on government intervention.

A female resident in Belleplaine, St. Andrew stated,

> Well, I would say, they get up and do things [themselves]. People that could help themselves [would do so]… But who can't, yes, [you] have to wait pun government.

This was also borne out over the course of the interview with the seventy-three-year-old female resident of St. Andrew. The question was asked about the relocation of her house after it was damaged during a flood and the erection of the new home in a different position on her plot of land,

> The family [take] it down [the damaged building]... [and we get this] from Urban [Development Commission]... [You] see when them [Urban] come and look, them say it can't build back there [hand gesturing to where her house previously stood]...So they would build it here and [you] would get help [in that way from them].

A similar experience occurred in Holetown, St. James with the female resident who is sixty-eight years old. She described the people involved in cleaning up after a flood brought mud and silt into her neighbourhood,

> Your own living area, [you] would, more than likely clean your own little area [because] it would be, there would be stones. There were lots and lots of stones along that road coming down by the church...Lots of stones...Just clean up, clean up the area. If there was mud up here in this...[hand gesturing] little courtyard area, [you would] hose it down, clean up. Just like when we had the volcanic [ashfall from Mount Soufriere in 2021] as much as we could, in our own area... until the government came through with the sandbags and the bags and filled up.

Resilience theory by Ungar (2018) and van Breda (2018) supports these findings that individuals and families had "multilevel coping strengths" and "protective resources" which were personal and arose out of social relationships (with family and friends) and were not only limited to government intervention. After their own resources have been fully utilized and extended, they acknowledged the need of government intervention, often to accomplish bigger tasks than they were able to do.

Limitations

One significant limitation was that the Belleplaine elderly participants demonstrated the success of personal resilience, effective macro-level policies and programmes and a supportive kinship network, all without the need of known social work activism. The explanation for this success was not within the framework of the theory of environmental justice. The focus of environmental justice is on the lack of corporate and governmental entities to facilitate equal access to resources, equal opportunities to be heard and equal implementation of action plans and programmes in all communities as needed. What it did not explain in this study is why flooding continues to be a severe problem in civic-resources-rich Holetown but not as severe in rural Belleplaine. It was initially believed that a focus on the theory of environmental justice offered possibilities in explaining the differences between the communities. Resilience theory proved a better fit.

Discussion

The study examined elderly residents residing in two communities in Barbados: Belleplaine, St. Andrew, which is located on the rural east coast, and Holetown, St. James, which is located on the urban corridor of the west coast of the island. The research questions examined the lived experience of elder persons living in Belleplaine, St. Andrew and Holetown, St. James and how they negotiated receiving assistance during the flood events. Two theories were utilized as the theoretical framework, resilience theory and environmental justice from a social work perspective. Three themes emerged: they have a trust in God, they know the geography of their communities, and they have an immediate, reliable social support network of relatives and friends as well as a reliance on the provision of the government regarding flood preparedness and recovery.

In the first theme, "they have trust in God," an example of this is a sixty-seven-year-old participant in Holetown who pointed to, what was for her, evidence of the salvation of the Lord in past flood events. Her survival of floods was attributed to His saving her household. This inspires her to continue to pray for His saving intervention. Resilience theorist Greene (2002) includes an older person's spirituality as an enabling factor, even as the elder reminisces and forms a cohesive reconstruction of past stressful events and the significance of her outcomes (Greene & Cohen, 2005). The result of this study maintains Goddard-Durant (2019) research where she found that the methods of handling overwhelming difficulties in the Barbadian middle class include drawing on Christian-based beliefs and accepting a community understanding of the circumstance with the conclusion that it was not insurmountable.

In the second theme, they know the geography of their communities, this was evidenced by the participants' knowledge of how water flowed around their community, from nearby high geographical areas towards the plain or the coast of their communities. One non-resident female, who had worked in Holetown for fifty years, watched the sky and predicted how the rain would "service" (affect) her worksite. Crate (2006) asserts that this elder's knowledge has a role to play in the community. Caquard et al. (2009) encourage this use of elder's relevant knowledge of the geography of their community and knowledge of the consequent expected flow of water in the local area. It precludes, for the participants in this study specifically, the need of supporting data from mapping technology.

In the last theme, they have an immediate, reliable social support network of relatives and friends as well as a reliance on the provision of the government regarding flood preparedness and recovery. Most participants had an immediate, reliable social support network of relatives and friends as well as a reliance on the provision of the government regarding flood preparedness and recovery.

In the Silva et al. (2019) findings, elderly Brazilian's showed a high capacity for resilience because of the combined strengths of financial autonomy, social support from their families and their exercise of religion. This is also in line with Lacet (2016), whose Haitian women combined the internal and external factors of family, community and agency supports, strong spirituality and an understanding of their role as determined leaders in their families.

These findings indicate a complete understanding emerging from theory, the previous literature and the research findings. That is the elderly is not an invulnerable social group, but they are also not as they are not incapacitated by their physical frailties or in a helpless state of dependency on macro-level interventions. The elderly of Belleplaine and Holetown demonstrate experience and local knowledge. Their local knowledge, their practise of faith and the reliability of their support systems cumulatively go beyond the single themes of a resilience experience. They instead create a synthesis of resilience more than the action of any one of these resources acting on its own.

The role of the social worker is to recognize the Belleplaine, St. Andrew's and Holetown, St. James' elderly experience of resilience against flooding. This phenomenological finding does not deny the evidence of past research and the environmental justice emphasis of social workers, Joseph and Doon (2022), which speak to the increasing physical frailty and additional health challenges and disabilities of elderly persons in Belleplaine, St. Andrew and Holetown, St. James and the need for their voices to be heard on environmental issues. However, the Barbadian experience in these communities demonstrates that the elderly have their own indigenous strengths and resources which cumulatively contribute to their ability to weather floods and maintain their social functioning. That is, the elderly's personal knowledge of their physical environment heavily contributes to their resilient response to flooding hazard.

Many of them express a general acceptance of the cumulative supportive roles of their families, neighbours and the government in the recovery process. Their belief in God's care and supervision prepares them to accept the oncoming threat of hydro-meteorological disturbances without overwhelming distress. Theirs is a successful range of outcomes, experienced and learned from over the course of their lives, in the face of a periodic flood risk to their communities. The social worker's role is to facilitate the promotion of this elder knowledge to the existing disaster management plans of the country and, in turn, encourage the use of flood preparedness tools provided by macro-agencies, by elderly persons and their caregivers.

Implications

More research is needed in local communities across the island, and in the contexts of other hazards, to have a generalizable definition of Barbadian elder resilience. Recognition of a Barbadian definition of elder resilience can be promoted in social policies, such as the Elderly Persons Act, to support the intent of government in maintaining an image of the Barbadian elder person to whom is due all possible social rights, well-being and security. The social worker's role is to facilitate the promotion of this elder knowledge to the existing disaster management plans of the country and, in turn, encourage the use of flood preparedness tools provided by macro-agencies, by elderly persons and their caregivers.

Conclusion

Elderly persons have a positive contribution to make towards the flood management policy in their communities. This social work research emphasized the strengths of the elder person in weathering the hazards of flooding in Belleplaine, St. Andrew and Holetown, St. James. They cumulatively lived eight lifetimes accumulating experiences of resilience, despite the lack of agency resources in Belleplaine and despite the severity of flooding in resource-rich Holetown. Their resilience methods involve the exercise of trust in God, knowing the geography of their communities and the reliability of the support of family and friends and lastly, of government intervention. These methods exist together in a synthesis of resilience against flooding that is unique to these communities' elderly residents.

This research was approved by the University of the West Indies Cave Hill/Barbados Ministry of Health Research Ethics Committee/ Institutional Review Board. No. No. 210504-B.

References

Babbie, E. R. (2010). *The practice of social research* (12th ed.). Wadsworth.

Beard, J. R., & Bloom, D. E. (2015). Towards a comprehensive public health response to population ageing. *Lancet, 385*(9968), 658–661. https://doi.org/10.1016%2FS0140-6736(14)61461-6

Beltran, R., Hacker, A., & Begun, S. (2016). Environmental justice is a social justice issue: Incorporating environmental justice into social work practice curricula. *Journal of Social Work Education*. https://doi.org/10.1080/10437797.2016.1215277

ben Hassen, H., Dghais, W., & Hamdi, B. (2019). An e-health system for monitoring elderly health based on Internet of Things and fog computing. *Health Information Science and Systems, 7*(1), 24. https://doi.org/10.1007%2Fs13755-019-0087-z

Berry, P., Clarke, K., Pajot, M., Hutton, D., & Verret, M. (2009). *The role of risk perception and health communication in adapting to the health impacts of climate change in Canada*. [report] Comprehensive Report prepared for the Climate Change Impacts and Adaptation Program, Earth Sciences Sector, Natural Resources Canada, Ottawa, Canada. https://doi.org/10.1007/978-94-007-0567-8_14

Biguenet, M., Chaumillon, E., Sabatier, P., Bastien, A., Geba, E., Arnaud, F., Coulombier, T., & Feuillet, N. (2023). Hurricane Irma: An unprecedented event over the last 3700 years? Geomorphological changes and sedimentological record in Codrington Lagoon, Barbuda. *Natural Hazards and Earth System Sciences, 23*(12), 3761–3788. https://nhess.copernicus.org/articles/23/3761/2023/nhess-23-3761-2023.html

Brockie, L., & Miller, E. (2017). Understanding older adults' resilience during the Brisbane floods: Social capital, life experience, and optimism. *Disaster Medicine and Public Health Preparedness, 11*(1), 72–79.

Brown, C. R. & Murphy, M. M. (2018). Barriers to informal caregiving of persons living with dementia in Barbados: A phenomenological inquiry. *West Indian Medical Journal, 67*(3). https://doi.org/10.7727/wimj.2017.142

Caquard, S., Pyne, S., Igloliorte, H., Mierins, K., Hayes, A., & Fraser Taylor, D. R. (2009). A "living" atlas for geospatial storytelling: The cybercartographic atlas of indigenous perspectives and knowledge of the Great Lakes region. *Cartographica: The International Journal for Geographic Information and Geovisualization, 44*(2), 83–100. https://www.researchgate.net

Chávez-Alvarado, R., & Sánchez-González, D. (2016). Vulnerable aging in flooded households and adaptation to climate change in cities in Latin America: The case of Monterrey. *Papeles Poblacion, 22*, 9–42. https://doi.org/10.22185/24487147.2016.90.033

Crate, S. (2006). Elder knowledge and sustainable livelihoods in Post-Soviet Russia: Finding dialogue across the generations. *Arctic Anthropology, 43*(1), 40–51. https://www.jstor.org

Cumberbatch, J., Drakes, C., Mackey, T., Nagdee, M., Wood, J., Degia, A. K., & Hinds, C. (2020). Social vulnerability index: Barbados–A case study. *Coastal Management, 48*(5), 505–526. https://www.researchgate.net

Department of Emergency Management. (2020). *Strategic plan 2019–2023—a more resilient Barbados*. Government of Barbados. https://dem.gov.bb

Dilby, L., Dickerson, S., Duff, M., & Vandermause, R. (2020). *Doing hermeneutic phenomenological research—a practical guide*. Sage Publications Limited.

Dhemba, J. (2012). Overcoming poverty in old age: Social security provision in Lesotho, South Africa and Zimbabwe revisited. *International Social Work, 56*(6), 816–827. https://doi.org/10.1177/0020872812446529

Dhemba, J., & Dhemba, B. (2015). Ageing and care of older persons in Southern Africa: Lesotho and Zimbabwe compared. *Social Work & Society, 13*(2). file:///C:/Users/mshoye/Downloads/salbus,+05+article_dhemba+dhemba%20(1).pdf

Devonish, D. & Nurse, L. (2023). The 2023–2028 national policy on ageing for Barbados—Making healthy and active ageing a reality for all. In *The Ministry of People Empowerment and Elder Affairs, Barbados.*

Donovan, S. & Harper, D. (2005). The geology of Barbados: A field guide. *Caribbean Journal of Earth Science, 38*, 21–33. https://www.mona.uwi.edu/

Downes, A., Cashman, A., Cumberbatch, J., & Lashley, J. (2015). *The Barbados National Report for the Third United Nations Conference on Housing and Sustainable Urban Development—Habitat III*. Cave Hill Campus, Barbados: University of the West Indies.

Evanson, D. (2014). *Country document for disaster risk reduction: Barbados, 2014*. Department of Emergency Management: Barbados.

Goddard-Durant, S. (2019). *"Resilience" in Barbados: Bein' uh Work in Progress.* [PhD Thesis]. University of Guelph. https://atrium.lib.uoguelph.ca

Greene, R. R. (2002). Holocaust survivors: A study in resilience. *Journal of Gerontological Social Work, 37*(1), 3–18. https://d1wqtxts1xzle7.cloudfront.net

Greene, R. R. & Cohen, H. L. (2005). Social work with older adults and their families: Changing practice paradigms. *Families in Society, 86*(3), 367–373. https://www.researchgate.net/

Greene, R., & Greene, N. (2022). A research-informed human behavior and social work practice approach: The resilience-enhancing stress model. *Journal of Evidence-Based Social Work, 19*(4), 401–407. https://doi.org/10.1080/26408066.2022.2075246

Groenewald, T. (2004). A phenomenological research design illustrated. *International Journal of Qualitative Methods, 3*(1), 42–55. https://doi.org/10.1177/160940690400300104

Hambleton, I., Clarke, K., Broome, H. L., Fraser, H. S., Brathwaite, F. & Hennis, A. (2005). Historical and current predictors of self-reported health status among elderly persons in Barbados." *Rev Panam Salud Publica*, *17*(5/6), 342–352. https://www.scielosp.org

International Labour Organisation. (2022). *Social protection expenditure review, Barbados*. ILO. https://www.ilo.org

Johnson, H., Ling, C., & McBee, E. (2014). Multidisciplinary care for the elderly in disasters: An integrative review. *Prehospital and Disaster Medicine, 30*(1), 72–79. https://doi.org/10.1017/S1049023X14001241

Joseph, D. D. & Doon, R. A. (2022). The social and economic implications of environmental justice for the elderly: A case for social work interventions in the Caribbean. In *Environment, Climate, and Social Justice* (pp. 17–43). Springer. https://doi.org/10.1007/978-981-19-1987-9_2

Kirton, M. (2013). *Caribbean regional disaster response and management mechanisms: Prospects and challenges*. Brookings-LSE Project on Internal Displacement.

Lacet, C. (2016). Fanm Vanyan: A cultural interpretation of resilience in Haitian women. *Journal of International Women's Studies, 17*(3), 45–58. https://vc.bridgew.edu/jiws/vol17/iss3/5

Limonta, M., Garzanti, E., Resentini, A., Andò, S., Boni, M., & Bechstädt, T. (2015). Multi-cyclic sediment transfer along and across convergent plate boundaries (Barbados, Lesser Antilles). *Basin Research, 27*(6), 696–713. https://doi.org/10.1111/bre.12095

Pan American Health Organization & International Telecommunication Union. (2023). *Decade of healthy aging in the Americas—situation and challenges—the role of digital technologies in aging and health*. PAHO. https://iris.paho.org/

Pietkiewicz, I., & Smith, J. A. (2014). A practical guide to using interpretative phenomenological analysis in qualitative research psychology. *Psychological Journal, 20*(1), 7–14. https://doi.org/10.14691/CPPJ.20.1.7

Rambaree, K. (2020). Environmental justice in the case of the Chagos marine protected area: Implications for international social work. *Sustainability, 12*, 8349. https://doi.org/10.3390/su12208349

Silva, E. G. D., Eulálio, M. D., Souto, R. Q., Santos, K. D., Melo, R. L., & Lacerda, A. R. (2019). The capacity for resilience and social support in the urban elderly. *Ciencia & Saude Coletiva, 24*, 7–16. https://www.scielo.br/j/csc/a/46r8zgKPzJZ978rsS4pj4bj/?format=pdf&lang=en

Springer, C. (2012). *Barbados national report on ageing 2007–2011*. https://issuu.com/barbadoslabourparty/docs/barbados_national_report_on_ageing_

Sullivan, M. P. (2017). *Hurricanes Irma and Maria: Impact on Caribbean countries and foreign territories*. Library of Congress Washington DC. https://apps.dtic.mil/sti/pdfs/AD1172110.pdf

Thomas, A., Baptiste, A., Martyr-Koller, R., Pringle, P., & Rhiney, K. (2020). *Climate Change and Small Island Developing States, 45*, 1–27. https://doi.org/10.1146/annurev-environ-012320-083355

Town and Country Development Planning Office. (2023). *Barbados physical development plan amendment—toward a green, prosperous, healthy and resilient nation*. Barbados. http://www.townplanning.gov.bb/pdp/Downloads/files/PDP-August-2023.pdf

Ungar, M. (2004). A Constructionist discourse on resilience: Multiple contexts, multiple realities among at-risk children and youth. *Youth & Society, 35*(3), 341–365. https://doi.org/10.1177/0044118X03257030

Ungar, M. (2018). Systemic resilience. *Ecology and Society, 23*(4). https://www.ecologyandsociety.org/vol23/iss4/art34/

van Breda, A. D. (2018). A critical review of resilience theory and its relevance for social work. *Social Work, 54*(1), 1–18. https://doi.org/10.15270/54-1-611

Climate Change-Induced Disasters in Manicaland Province in Zimbabwe: The Need for Continuous Professional Development

Molly Shayamano, Poppy Masinga, and Nkosiyazi Dube

Climate change has emerged as one of the most pressing challenges worldwide, causing an increase in the intensity, occurrence, and frequency of meteorological disasters. Such disasters include floods, drought, wildfires, cyclones, and heatwaves, which have extensive consequences for individuals, families, groups, and communities, particularly among vulnerable populations. Climate change-induced disasters often stem from a rise

M. Shayamano (✉) · N. Dube
University of the Witwatersrand, Johannesburg, South Africa
e-mail: mollyshayamano@gmail.com

N. Dube
e-mail: Nkosiyazi.Dube@wits.ac.za

P. Masinga
South African College of Applied Psychology, Cape Town, South Africa
e-mail: poppy@sacap.edu.za

D. D. Joseph et al. (eds.), *Climate Change*,
https://doi.org/10.1007/978-3-031-83037-2_5

of greenhouse gases in the atmosphere, as well as extreme weather conditions (Dembedza et al., 2023), and have accounted for about 84% of natural disasters globally in the period between 2008 and 2017 (International Federation of Red Cross and Crescent Societies [IFRC], 2018). In the same period, approximately two billion people were directly affected by weather-related disasters, where the most disadvantaged individuals live in the Global South, where they suffered immensely from the effects of these disasters (IFRC, 2018).

Dominelli (2011) opines that social workers can play a significant role in the climate change and disasters debate, yet the profession has not been actively participating in these debates. There is a need for social work educators to incorporate environmental justice issues from a climate change perspective, along with practice skills, into the social work curriculum. This allows practitioners to learn about climate change and disasters, enabling them to arise to address the challenges posed by climate change.

Despite growing recognition of the importance of social workers in disaster situations globally, Mhlanga et al., (2019); Shokane (2019); and Muzingili (2016) noted that social workers' limited exposure to a disaster-oriented curriculum negatively impacts their efforts. Most social workers were exposed to a generic curriculum that is silent on disaster issues (Hay & Pascoe, 2021), yet some of them end up working in organisations that respond to disaster events, where they need to be at the forefront. Equipping social workers with those skills relevant for disaster intervention during training is fundamental. The social work profession has a primary mandate of protecting people's well-being (Harms & Alstorn, 2018), thereby making practitioners important role players in the disaster management field, as they provide mental health support, immediate relief, and resilience plans to the disaster victims and survivors. Some social work training institutions have started to mainstream disaster-informed practice in the qualifying social work programmes (Dominelli, 2023), which benefits the social workers who are still in training. Continuous professional development (CPD) stands as a critical tool for empowering seasoned social work practitioners who lack disaster management training with specialised knowledge and skills (Dziro, 2013; Hudson et al., 2021). This is needed to effectively navigate the complexities of disaster preparedness, prevention, response, and recovery. In this regard, social work practitioners that are equipped with expertise in disaster related research, crisis intervention, risk assessment, disaster planning, resource mobilisation, and trauma-informed care, will have a significant positive impact towards the delivery of disaster-oriented social work practice.

This chapter examines gaps in social work training pertaining to environmental justice when working in disasters and discusses the relevance of the social work profession prior to, during, and in the aftermath of disaster events. The experiences of social work practitioners in Manicaland Province were examined by employing a qualitative research method, gathering data from social work practitioners through focus group discussions. The chapter commences by delineating the broader context of climate change and its resultant disasters across Africa, setting the stage for a focused exploration. It then presents a detailed examination of climate change-induced disasters within Manicaland Province, providing a critical backdrop for understanding the challenges faced by the participants. In doing so, this chapter enriches the discourse on the necessity and implementation of disaster-focused social work education, training, practice, and scholarly research, thereby offering valuable insights into adaptive strategies and interventions in the face of climate adversities.

Climate Change and Africa

When considering the rapidly changing climate worldwide, it is notable that Africa is the most affected continent, situated as it is primarily within the tropics, and characterised for this reason by high temperatures and significant seasonal rainfall variations. Currently, Africa's capacity to cope with the devastating effects of climate change remains severely hampered (The African Development Bank, 2024). Given the continent's high climate change exposure and limited capacity, its communities are severely affected. In 2022, approximately 110 million people in Africa were directly impacted by climate change and related disasters (World Meteorological Organization [WMO], 2023). These catastrophic events caused property damage, livelihood destruction, displacement, loss of lives, and heightened health risks. Factors contributing to Africa's vulnerability include long-term temperature anomalies, rising sea levels, and erosion. The most frequently observed disasters in Africa are droughts, cyclones, floods, and heatwaves (WMO, 2023). Jayawardena (2015) opines that floods result from excessive rainfall, while droughts occur due to low rainfall. The WMO (2024) conceptualises a cyclone as involving intense rotating storms, which originate from tropical oceans, characterised by strong wind speed, and heavy rains (WMO, 2024).

The Southern Africa region also bears the brunt of climate change and related disasters, which threaten its communities and ecosystem. Between 2000 and 2023, the region experienced an increase in high-impact tropical cyclones and flooding, which caused tremendous damage to the ecosystem (Tevera & Raimundo, 2021). For example, the 2022 floods in KwaZulu-Natal and the Eastern Cape provinces of South Africa affected 121,687 people and resulted in the loss of 435 lives, infrastructure damage, deaths, illnesses, injuries, emotional trauma, and vegetation damage (Bouchard et al., 2023). A significant number of households that are dependent on small to medium enterprises (SMEs) were left more vulnerable as their main sources of income were obliterated by the heavy rains. For example, SMEs such as market vendors and local shops experienced severe disruptions in their businesses as the heavy floods washed away their stocks and destroyed infrastructure.

In 2019, Mozambique was affected by two severe cyclones, six weeks apart, affecting 1.7 million people (World Health Organisation [WHO], 2021). The two severe cyclones left a trail of destruction in Mozambique, as well as the numerous meteorological disasters affecting the ecosystem. Mozambique's extreme weather conditions exacerbated by climate change have become more common, resulting in tropical storms and cyclones, which cause extreme damage to the environment, people, animals, and vegetation.

In Zimbabwe, the most prevalent weather-related disasters are droughts, floods, and cyclones. The frequency and occurrence of cyclone incidents in the country have intensified (Dembedza et al., 2023). In more recent years, Zimbabwe experienced the devastating effects of Cyclone Idai in March 2019, Cyclone Chalane from December 2020 to January 2021, Cyclone Eloise in January 2021, Tropical Storm Ana in January 2022, and Cyclone Freddy in March 2023. These disasters caused extreme flooding, dam collapses, mudslides (Mukurazhizha & Matanga, 2023), strong winds, heavy rains (Mavhura & Aryal, 2022), road damage, and bridge collapse (Chatiza, 2019), which resulted in livelihood alterations, displacement, deaths, illnesses, injuries, and destruction of property, houses, and vegetation.

In addition, climate change-related disasters threaten people's health and well-being. The occurrence of extreme weather events such as floods, heatwaves, droughts, and cyclones has been linked to various health issues such as cardiovascular diseases, respiratory diseases, mental health, malnutrition, and infectious diseases such as cholera, caused by drinking contaminated water (Ghazali et al., 2018; World Economic Forum, 2024). During Cyclone Idai in Zimbabwe, many people in Manicaland Province experienced health problems, which resulted in 21, 952 cyclone-related hospital admissions (Chatiza, 2019). In most instances, disasters destroy the agricultural sector, resulting in food insecurity issues in the affected areas, thereby increasing risks of malnutrition to the disaster victims (Chatiza, 2019). The World Economic Forum (2024) asserts that disaster events also have indirect and long-term effects on people's health and well-being, such as stunted growth in children, and anxiety disorders.

Given the increase in the occurrence of climate change-related weather disasters, the social work profession needs to deepen its engagement in environmental issues, since the practitioners work more closely with disaster victims and survivors. Many studies have documented the critical role played by social workers in the face of disaster events (Dominelli, 2018; Machimbidza et al., 2022; Mhlanga et al., 2019; Rapeli, 2017), making social work practitioners crucial members of the disaster management team. Mhlanga et al. (2019) note that disaster occurrences are linked to community vulnerability, providing a foundation within which social workers can demonstrate their capabilities in disaster-oriented social work practice.

An Overview of Climate Change-Induced Disasters in Manicaland Province

Zimbabwe is a landlocked country, situated in the Southern parts of the African continent. As of 2022, Zimbabwe has a total population of 15, 178, 957 people, with a land area of 390,757 square kilometres (Zimbabwe National Statistics Agency [ZimStat], 2023). The country's economy is agro-based (Risiro, 2019), and the occurrence of climate change-induced disasters is destroying the country's economy by damaging the agriculture sector, due to flooding and drought. This study was conducted in Manicaland Province which is situated in the eastern part of the country, and is inundated by meteorological disasters, particularly cyclones and tropical storms. According to the census 2022 results, Manicaland Province is the second largest populated province in Zimbabwe, accounting for 13.4% of the total population (ZimStat, 2023). The Province is characterised by ten districts, namely Makoni,

Buhera, Chimanimani, Chipinge, Mutare, Mutasa, Nyanga, Mutare City Council, Chipinge Town Council, and Rusape Town Council. Chipinge and Chimanimani are the most disaster-prone districts within Manicaland Province.

Since 2000, the Province has been affected by about eight meteorological disasters. Cyclone Eline hit the Province and other parts of the country in February 2000, causing heavy rains and strong winds, which resulted in extensive damage (Agritex et al., 2000). The cyclone impacted the lives of many people due to its high magnitude. In total, 540,000 households were affected, which translates to approximately 2.7 million people (Mavhura, 2020). These people were exposed to diseases, loss of livelihood, emotional trauma, and displacement.

Two districts in Manicaland Province, namely Chipinge and Chimanimani, were severely affected. In Chimanimani, about 1050 households were directly affected by the cyclone, and many people were displaced. Flooding induced by the cyclone caused multiple waterborne diseases, with diarrhoea and malaria being the deadliest (Agritex et al., 2000). The impact of these disasters heightened social problems, such as food insecurity and homelessness, exacerbating existing vulnerabilities within the affected communities. In this instance, social work practitioners should intervene and address such problems and support disaster victims and survivors in their recovery process by providing mental health support and linking people with resources.

In February 2000, there were 2737 reported cases of malaria, and 406 individuals suffering from diarrhoea in Manicaland (Agritex et al., 2000). The Province is characterised by an agrarian economy, and many households depend on agricultural produce for survival, and as a source of income. Cyclone Eline destroyed the livelihoods of many a household. In such situations, social workers could assist families in accessing immediate financial relief, such as food relief, and cash transfers. Additionally, social work practitioners could facilitate community-based initiatives in the affected communities to enhance livelihood opportunities and assist disaster victims in building support networks to regain their economic stability.

Furthermore, a total of 6872 livestock and 1072 hectares of crops were washed away by floods induced by Cyclone Eline, thereby affecting food security in the Province (Agritex et al., 2000). Infrastructure was also damaged, with about 13 healthcare facilities, roads, and schools being partially destroyed. Social work practitioners in collaboration with other emergency response teams could assist in rebuilding communities' essential infrastructure in their recovery plans. The practitioners may also play a crucial role in connecting affected communities with organisations and government programmes offering reconstruction services, repairs, and other essential needs. The social workers' long-term plans could involve advocating for funding to rebuild infrastructure and resuscitate the affected healthcare systems.

In 2017, Cyclone Dineo, a Category 3 cyclone, landed in Zimbabwe affecting about 4131 households, 192 people lost their lives with approximately 136 individuals suffering injuries (Mavhura, 2020). About 20,000 homes were destroyed and 130,000 individuals were victims of the cyclone in Manicaland (Matutu, 2021), leaving many people homeless and vulnerable. Under these circumstances, the social work practitioners could link the affected individuals with organisations offering medical assistance, housing, and immediate relief services. Mental health support is also crucial, supporting individuals who could have lost their loved ones, and all their possessions.

After Cyclone Dineo, the Province was hit by Cyclone Idai in March 2019, which tragically followed the same pathway as the previous cyclone, directly affecting 52,027 households (Mavhura, 2020). Flooding induced by Cyclone Idai resulted in 347 deaths, while 200 individuals were injured, approximately 344 people went missing (Mavhura, 2020), and about 51,000 individuals were displaced (Chatiza, 2019). Social workers' services are critical in such situations to support disaster victims through psychosocial support and linking victims with organisations and government programmes offering immediate relief services, such as food aid and temporary shelter. However, the social workers' response should go beyond immediate relief and should encompass long-term and sustainable community recovery plans. Manicaland Province was the epicentre of Cyclone Dineo in the country, with heavy rains, flooding, and strong winds leaving a trail of destruction in their wake. As a result, people were left more vulnerable, and in urgent need of emergency assistance.

In addition, flooding caused by Cyclone Idai destroyed 17,608 homes, 12 healthcare facilities, and 139 schools (IFRC, 2020), where about 43 schools were forced to close temporarily, which affected 9084 learners. The social work practitioners, as part of a multidisciplinary team in disaster management, have a critical role to play in advocating for the rebuilding of the destructed infrastructure. The agricultural sector was also severely affected by the Cyclone, as it wiped out more than half of the land that had maize crops, tuners, and banana plantations (IFRC, 2020). Approximately 18 irrigation schemes in the Province were destroyed, and 17,876 livestock were lost. Infrastructure damage was also severe, with more than 90% of road networks being destroyed and many bridges being totally damaged. As a result, disaster response initiatives were derailed, as the disaster response team struggled to access the affected areas due to damaged road networks. Approximately six weeks after Cyclone Idai, Cyclone Kenneth hit the same Province, while people were still struggling to recover from the deadly destruction of Cyclone Idai (Dembedza et al., 2023). The Cyclone also left a trail of destruction in the Province, worsening the well-being of the inhabitants.

In December 2020, Cyclone Chalane made landfall in Zimbabwe, with flooding and strong winds in Manicaland Province, particularly in Chimanimani and Chipinge Districts (Office for the Coordination of Humanitarian Affairs [OCHA], 2020). Chalane landed in Zimbabwe as a tropical depression. As a result, minimum damage was recorded in the Province. However, the strong winds blew off house roofs and disrupted internet connectivity, cellular networks, and electricity. A few amenities were mildly damaged, these included Chimanimani Hospital, a church, and a local school. OCHA (2020) notes that some families, who were internally displaced due to Cyclone Idai, were still living in temporary shelters when Chalane struck, and they had to be evacuated again to safer areas. Social work practitioners ought to be active participants in disaster management structures, offering their expertise in disaster response and recovery plans. In this instance, the practitioners could assist in the evacuation process, mobilising resources and building social support networks for the disaster victims. Being affected by two disasters within a short period can be detrimental for the affected, thereby necessitating for social workers to offer crisis intervention and mental health support to the victims. The World Economic Forum (2024) asserts that disasters cause mental health issues such as anxiety, trauma, and post-traumatic stress disorder. Thus, social workers' interventions could help disaster victims in managing their emotions.

In January 2021, Cyclone Eloise, yet another climate change-induced weather disaster hit Manicaland Province. There was a short period between Chalane and Cyclone Eloise; hence, the Province had not yet recovered from the damages caused by the preceding disaster. Like previous cyclones, Cyclone Eloise was characterised by heavy rains and very strong wind speeds, which damaged the environment and infrastructure that had already been weakened by previous cyclones (IFRC, 2021). More bridges, roads, and dam walls collapsed, leading to flash floods, and posing a health threat to people and animals. Unlike other cyclones, Cyclone Eloise was not declared a national disaster (IFRC, 2021), but it caused significant damage and affected many people in the affected areas, approximately 1889 individuals (IFRC, 2021). Field crops, people's dwellings, and livestock were destroyed. Three individuals lost their lives, where 15 houses and three schools were destroyed in Manicaland.

In addition, 265 people in the Province had to be evacuated to safer areas. These people faced double tragedy, as Covid-19 was still a threat. In the evacuation centres, people faced a heightened risk of contracting Covid-19, since maintaining social distancing was a challenge. The 2021 IFRC report stated that four individuals tested positive for Covid-19 at one of the evacuation centres in Manicaland Province during this period. In cases like these, social work practitioners should promote health education by providing disaster victims with information on disease prevention and ensuring access to health care.

In late January 2022, Zimbabwe experienced Tropical Storm Ana, which also brought heavy rains, causing flash floods that led to significant damage to infrastructure and the environment. Approximately 845 houses and 51 primary and secondary schools were damaged (OCHA, 2022). In Manicaland, the storm affected 1400 individuals (e.g. 282 households). Infrastructure damage was also witnessed in the Province, with 13 schools, nine bridges, and various road networks being destroyed. Another cyclone landed in Zimbabwe in the early months of 2023. Cyclone Freddy caused two deaths in Manicaland (Buwerimwe, 2023). The strong winds associated with the cyclone blew off the roofs of several houses in the Province. The Metrological Services Department reported that seven cattle were struck by lightning.

The effects of climate change and related disasters in Manicaland Province, Zimbabwe, are undeniable. The frequency and severity of these weather-related events, such as floods and cyclones, have had several implications for people's livelihoods, health, life, and overall well-being throughout the Province, where infrastructures such as houses, schools, healthcare facilities, and roads have been severely damaged. In the face of such adversity, the role of social work professionals emerges as crucial to enhancing resilience and adaptation. As communities in Manicaland grapple with the consequences of climate change, disaster-oriented social work training, practice, and research become increasingly necessary.

Overview of Disaster-Oriented Social Work Training, Practice, and Research

Disasters continuously affect both the environment and its inhabitants. The occurrence of these disasters poses new and complex challenges for social work practitioners working directly with affected individuals, families, groups, and communities. As such, social workers play a critical role in the disaster management discourse. Disaster-oriented social work is a specialised field that specifically focuses on disaster-related issues that harm people and the environment (Pyles, 2017), offering support to disaster victims and survivors through engaging in disaster preparedness, developing disaster recovery plans, and building community resilience. Dominelli (2014) believes that generic social work skills, such as active listening, communication, networking, advocacy, problem-solving, and advocacy apply to all fields of practice, including disaster-oriented social work practice. While generic social work skills apply to all fields of social work practice, disaster-oriented social work practitioners need to be equipped with crisis intervention, trauma response, counselling, planning, networking, and collaborating skills since they practice in high-stress environments. However, the visibility of social work practitioners remains limited (Mhlanga et al., 2019) due to a lack of exposure to a disaster curriculum during training (Nyahunda et al., 2019).

There is an absence of a disaster-oriented social work curriculum that demonstrates social workers' relevance during disaster situations. Dominelli (2023) is of the opinion that social work programmes across the globe have not been covering disaster interventions. The TATA Institute of Social Sciences in India stands out as the only training institution that initiated disaster-oriented social work training at a time when others had not yet embraced this practice (Dominelli, 2023). This initiative dates back as early as the 1940s, marking an early commitment to equipping social workers with relevant skills and knowledge for disaster practice.

It is, therefore, fundamental to address the deficit of practice-related skills on disaster management in social work training, where disaster-related content in existing social work modules, or by developing a standalone disaster management module that adequately prepares practitioners to engage in disaster issues, effectively. To accelerate disaster-oriented social work, Rambaree (2020) recommends incorporating environmental social work education into social work students' training consistently over their four years of training, rather than adding environment-related content as a unit, or as single module. This capacitates practitioners to assume disaster management roles with integrity.

In practice settings, social work professionals have long held critical roles during the aftermath of natural disasters to assist disaster victims and survivors (Zakour, 1997); however, their involvement has always been limited in scope, usually only providing counselling and humanitarian assistance. Hay and Pascoe (2021) concur that practitioners are more engaged in the response and recovery phases. Thus, the practitioners' interventions have not been holistic, as they were not actively participating in other phases of disaster management, namely prevention and preparedness. For social work interventions to be effective, the practitioners ought to be actively involved in preparing for, preventing, responding to, and recovering from disasters. It is therefore critical for the social work curriculum to prepare social workers for the advent of disasters, including participating in policy debates and disaster management discussions.

In cases where the curriculum fails to prepare social workers for disaster practice, and knowledge is limited, Hay et al. (2021) recommend engagement with a broader literature to understand social workers' roles and interventions during disasters. However, this is a challenge for disaster social work, since such literature remains relatively scarce (Mhlanga et al., 2019). Most studies have documented literature on social work interventions in the response and recovery phases, with limited information on social work disaster prevention and preparedness. Recently, publications on social work interventions in emergencies and disasters have increased (Alston et al., 2019; Dominelli, 2016, 2023; Matlakala et al,. 2022; Mhlanga et al., 2019; Rapeli, 2017) publishing extensively on disaster social work practice. Practitioners may rely on the published literature to build their knowledge in disaster-oriented practice. However, social work practitioners may not do this, resulting in the need for continuing professional development.

Furthermore, in Australia, researchers from Monash University and the University of Melbourne established a Social Work Disaster Network to lead disaster social work research and conduct cutting-edge research by means of which to generate new knowledge on social work interventions, responses, as well as just policies, and practices (Monash University,

2016). This constitutes a positive endeavour to spearhead research on the social work profession and disaster practice, which is likely to increase practitioners' knowledge of disasters. In the context of disaster-focused social work, continuous professional development proves crucial, as it has the potential for increased practitioner knowledge about disasters and empowers practitioners to remain competent in practice, ultimately enhancing the practitioners' capacity to support individuals, families, groups, and communities during times of crises.

Continuous Professional Development

Continuous professional development refers to the education or advancement of knowledge by professionals after completing formal training (Halton et al., 2015) to enhance skills and improve practice. Senge (1990) confirms that people should continually expand their knowledge and capacity to deliver effective results in an ever-changing environment. Halton et al. (2015) believe that initial education or training is not sufficient to last a lifetime for professionals, which prompts the need for lifelong learning. Given the dynamic nature of the environment, social work practitioners ought to adapt and develop themselves to remain relevant.

In addition, continuous professional development takes different forms, such as formal education and training, professional supervision, work-based learning, self-directed learning, online learning, reflective practice and feedback, professional conferences, peer learning, and networking (Research in Practice, 2023). All these CPD types are avenues via which social work practitioners can enhance their knowledge in disaster-oriented practice. In most countries, social work is a regulated profession, and practitioners are required to record their CPD activities when renewing their practice license annually (Halton et al., 2015). Table 5.1 shows the different types of CPD for social workers.

Participation of social work practitioners in any of the above CPD training programmes is critical to addressing the deficiencies in practice. The social work curriculum in many training institutions has often neglected the inclusion of disaster issues. As a result, many practitioners failed to engage effectively in disaster practice. This can be addressed if practitioners prioritise CPD on disaster-oriented social work practice. Through CPD, the practitioners will gain relevant knowledge and skills that prepare them for disaster practice, which can also be disseminated

Table 5.1 CPD for social workers—adapted from Halton et al. (2015) and Research in Practice (2023)

Continuous professional development type	*Description*
Formal education and training	• This includes attending formal academic courses such as postgraduate diplomas and postgraduate degrees offered by educational institutions such as colleges and universities
Self-directed learning	• This involves reading articles, reviewing books, book chapters and articles, and using the internet to update knowledge
Professional supervision	• This relates to professionals accessing structured support and guidance from an experienced professional to support growth
Professional conferences, seminars, and webinars	• It involves attending and/or presenting at conferences, symposiums, and seminars allowing the practitioners to stay updated with the latest developments and trends in their fields of practice • It also includes involvement in a professional body
Work-based learning	• This involves updating knowledge or learning through engagements such as in-service training, case studies, and work shadowing

to other professionals through knowledge-sharing platforms, such as seminars and colloquia.

Methodology

The qualitative research method for this study was constituted by focus group discussions with social work practitioners in Manicaland Province, Zimbabwe. A total of three focus group discussions were conducted, with 26 social work practitioners working for the government (Department of Social Development) and non-governmental organisations in Manicaland Province, Zimbabwe. Purposive sampling was utilised to select the 26 social work practitioners for this study, ensuring they possessed the relevant knowledge and experience on disaster-oriented social work practice in the country. The practitioners were recruited through online platforms by posting calls for participation on social workers' professional forums. Ethical approval was granted by the Ethics Committee of the University of the Witwatersrand (07 June 2021/No. H21/04/35). In addition, informed consent to participate in the study was obtained from the social work practitioners.

The inclusion criteria to participate in the study included an identified social worker with either a certificate, diploma, honour's, Masters, or PhD. The participant ought to be employed and exposed to a disaster management experience in the field. Table 5.2 shows the demographic characteristics of the study participants by race, age, gender, qualification, and years of work experience.

Data Collection

An interview guide was created based on the study objectives. The participants were asked various questions related to climate change-related disasters, including their understanding of such disasters, the integration of climate change into social work practice, the disasters most common in their locality, their role in addressing such events, the necessary skills for effective disaster intervention, the topics taught during training, barriers for intervention suggestions for improvement, and their suggestions for enhancing disaster social work training and practice.

The focus group discussions were 45–60 minutes long, and a recorder was used to record the focus group discussion sessions. Thematic analysis was employed to analyse the gathered data following Braun and Clarke's (2006) six-phase data analysis framework, during which main themes were generated. The six phases involve familiarisation with gathered data, producing initial codes to organise the data into meaningful groups; searching for potential themes; reviewing and refining the potential themes; naming the themes; and lastly, writing a report (Braun & Clarke, 2012). The data analysis process involved transcribing recordings, continuous reading of the transcripts, and jotting down important notes, labelling and organising data into meaningful groups, developing a thematic table, and ensuring that each theme has sufficient data, refining themes. Finally, it involved discussing the themes and how these fit into the broader study. The participants were cited verbatim in what appears below.

Table 5.2 Race, age, gender, highest educational qualification, and years of service in a disaster management context of the study participants

Race and Nationality	*Number of participants*
Black Zimbabweans	26
Total	26
Age	*Number of participants*
25–30	11
31–35	6
36–40	4
41–45	2
46–50	2
51 and above	1
Total	26
Gender	*Number of participants*
Males	8
Females	18
Total	26
Highest level of qualification	*Number of participants*
Certificate	1
Diploma	0
Honours degree	10
Master's degree	15
PhD	0
Total	26
Years of experience in disaster contexts	*Number of participants*
0–2 years	2
3–5 years	21
6–8 years	2
9–11 years	1
12–14 years	0
15 years and more	0
Total	26

Data Analysis

Theme One: Social Work Curriculum and Social Work Practice in Manicaland

The participants were all qualified and registered social work practitioners. All 26 participants received their social work training in Zimbabwe, with the majority indicating that they were not trained on disaster-related matters, citing this as the reason why they were failing to find their ground on disaster management issues. The practitioners' knowledge about climate change and disasters is therefore limited, which in turn affects their interventions in disaster prevention, preparedness, response, and recovery. Key insights from the study show that practitioners lack exposure to a disaster-oriented curriculum. In one of the focus group discussions, a practitioner stated, *"I was also never exposed to any disaster management module or topic during my training…"*.

Another practitioner commented, *"I graduated with an undergraduate degree in 2017, and within the four years that I was at university, I was not taught anything related to disaster management."* Another social work practitioner shared their perspective on the social work curriculum, stating, *"When I did my Bachelor of Social Work degree, we were not taught about climate change or disaster management…"*.

The above responses from the practitioners reflect the common challenge of gaps in the social work curriculum in preparing practitioners to take up roles in disaster issues in the field (Mhlanga et al., 2019; Murenje & Porter, 2020), despite the ever-increasing intensity and frequency of such disasters and their impact. The practitioners in the present study confirmed that there had been an increase in climate change-related disasters in Manicaland Province over the years. One of the practitioners highlighted, *"…in Manicaland, disasters occur regularly, especially cyclones* […] *a lot of cyclones have affected the Province* […] *and as social workers, we will be at the forefront in assisting the victims."*

Practitioners acknowledged their knowledge deficiency about disaster interventions. As such, most gained disaster knowledge in the field. One practitioner shared her experience by saying, *"I had to train myself on how to respond to disaster situations."* Another practitioner emphasised, *"I only got to know about disaster management and climate change when I started working* […] *everything that I now know regarding disasters I learned during practice, when I was already assisting victims of disasters."* Another social worker stated, *"I started knowing about Disaster Management in practice, since I work in a disaster-prone Province."* These sentiments highlight the gaps in social work training, which places practitioners at a disadvantage, as it affects their efficiency in disaster practice. In addressing this gap, many scholars have stressed the need to mainstream environmental issues in the social work curricula (Dominelli, 2016; Maripe & Sinkamba, 2014; Nyahunda et al., 2019).

Furthermore, some practitioners were privileged to receive training in disaster social work, although these are few. One of the practitioners commented, *"Yes, disaster management was part of the curriculum during training, we were taught about disaster management in the final year of study."* Another social worker shared her perspective on the curriculum issue, stating, *"Disaster management was one of the modules taught in the final year of my Social Work degree which was done in the first semester only* [...] *it was not at a broader scale."* These practitioners were at an advantage in practice, as they possessed some disaster knowledge gained from social work training. This proves that disaster-oriented social work training is critical in preparing practitioners to be competent in making a positive impact on disaster victims and survivors.

In addition, some practitioners learned about disaster social work during experiential learning in their third year of the Bachelor of Social Work degree, having been placed at organisations that respond to disasters. One of the practitioners shared: "*...the only time I got to know about social work roles during disasters was when I was doing my attachment at the Department of Social Development and there was a drought in the area that I was working in. So, we had to assist the community members who were affected by the drought... we gave them bags of maize."* The practitioner expressed that their interventions were only limited to the provision of food relief. This shows that they were not exposed to social work interventions in all four phases of disaster management, focusing instead on disaster response through providing immediate relief to disaster victims and survivors. For this reason, they were not adequately equipped for disaster practice, which affected the profession's effectiveness and recognition in disaster contexts.

The response from practitioners presented above proves the need for a CPD for social workers to enhance their knowledge of disaster interventions.

Theme Two: Visibility of Social Work Profession in Disaster Practice

Social work practice is relevant during disaster situations, as practitioners are required to intervene to assist disaster survivors in returning to their normal functioning and rebuild their communities. The study participants were practising in Manicaland Province, and thus strategically positioned to effectively contribute to disaster practice. However, their visibility proves insufficient, as they are mainly confined to providing immediate relief to disaster victims and survivors, with limited engagements in the spheres of prevention and preparedness. Visibility of the social work practitioners in the disaster management sphere serves as a critical starting point for the recognition of the profession. In this context, some subthemes shedding light on the diverse aspects of visibility are discussed. These include media presence and representation, inadequate involvement, and skill-based recognition.

Media Presence and Representation

Media presence and representation is critical to ensuring visibility of the social work profession's contributions to disaster management issues. However, this study found social work practitioners' visibility in the media (television, radio, newspapers, and online platforms) to be absent, thereby negatively impacting their recognition in disaster practice in Zimbabwe. Usually, the media reports on the activities of doctors and other professionals, yet little gets reported on social work activities and contributions in emergencies. Responses from social work practitioners support this during focus group discussions, with one participant arguing that "*in my experience, as social workers, we are not featured in media coverage of disasters. The media tend to focus on what NGOs provide, neglecting to show the role social workers play in supporting disaster victims and survivors.*"

In addition, another practitioner commented, "*It is disheartening to see our efforts downplayed. We will be in the community disaster events, providing services to the affected. However, the narratives in the media tend to focus on other responders, and little to nothing is mentioned about social workers.*" Both comments from practitioners highlight the lack of media representation of social work interventions in times of disasters. As a result, the practitioners' efforts often go unnoticed. As such, social work practitioners in Zimbabwe could possibly market their contributions in emergencies for other disaster management professionals and the community to recognise their importance. In Australia, for example, similar issues were reported by Hay and Pascoe (2021), arguing that media publications do not mention social work practitioners' contributions to disaster practice. Rasool (2020) concurs that social workers' professional capabilities are not utilised, and their efforts are go largely unacknowledged.

Inadequate Involvement

Most participants' responses signalled that social work practitioners become more engaged when disasters strike and in recovery stages to help communities return to normalcy. Limited social work engagements in the prevention and preparedness phases make their disaster-focused practice inadequate. This contributes to their restricted visibility in disaster management issues. One of the practitioners in the focus group discussions commented, "*We respond when disasters occur; for example, we provide counselling to those affected by the disasters and their families so that they can go back to their normal functioning.*" Another practitioner stated, "*Psychosocial support is provided to the affected victims to help them deal with the trauma associated with disasters.*"

In another focus group discussion, a practitioner commented, "*When disasters occur, we mainly offer food handouts and cash transfers to the victims of the disasters. These handouts will be from either the government or some humanitarian aid organisations* [...] *As social workers, we will be handing out social assistance to the affected people.*" The foregoing statements imply that practitioners are more actively involved in assisting those affected by disasters after disasters have already occurred by providing immediate relief, such as psychosocial support and food handouts to the disaster victims. This practice is not unique to practitioners in Zimbabwe, where Hay and Pascoe's study (2021) found that social workers in New Zealand are visible in activities related to immediate response and recovery, engaging in activities such as trauma-focused work, casework with new and existing social service clients, and aid distribution. While social work practitioners are relevant to disaster practice, they must expand their visibility by working with other disaster management actors in all aspects of disaster management.

Skill-Based Recognition

The skills demonstrated by practitioners are crucial in enhancing their visibility in disaster practice. Dominelli (2014) argues that social workers possess a range of skills and knowledge that can be tapped into when responding to disasters. Such skills include trauma-counselling, resource mobilisation, community organisation, and crisis intervention. Utilising these skills can scale up the recognition of the practitioners as key actors in disaster management. However, Hay and Pascoe (2021) believe that general social work skills are inadequate when it comes to disaster practice and recommend upskilling for improved disaster interventions. Social work participants also shared the same sentiments, with one practitioner commenting: *"...social work skills are applicable during disasters, but it is not enough because practising during disasters is complex."* The complexity of disasters has limited practitioners' effective engagement in disaster prevention and preparedness. Scaling up social work knowledge in disaster practice has the potential to enhance the practitioners' proficiency in dealing with complex and dynamic situations, such as disaster management. This will heighten their visibility and foster broader recognition of the profession in emergencies.

Theme Three: The Social Work CPD Training Programme

This study found that there is limited disaster training for social work practitioners, which negatively impacts their effectiveness. There is consequently a need for CPD. Social workers in the study emphasised the significance of disasters and climate change courses in social work training in preparing practitioners for disaster practice.

One practitioner shared, *"I believe that climate change and its related disasters is a contemporary issue, and since social workers work closely with communities affected by these disasters there is a need for them to be exposed to such issues during training."* This response underlines the importance of the inclusion of disaster issues in the curriculum, yet many practitioners were not exposed to disaster-oriented social work training.

The practitioners showed concern for these gaps in professional social work training. To this effect, the practitioners are opting to pursue further training to advance their knowledge and upskill. One of the practitioners expressed, "*... but I cannot safely say that I am knowledgeable about disaster management, so I will do a course in disaster management to enhance my knowledge.*" Another practitioner echoed the same sentiment, stating, *"If I had received formal training on disaster management, it would have made my job easier* [...] *a disaster management course is necessary."* This shows that the practitioners need training, resulting in the development of CPD as a training programme.

Discussion of Findings

The practitioners reported that the social work curriculum to which they were exposed remained silent on disaster management issues, leaving them ill-equipped to practice in emergencies efficiently. Findings from the study reveal that most social work practitioners were exposed to disaster-oriented social work in practice settings. Several studies reveal significant gaps in social work training on environmental issues (Huang et al., 2013; Matlakala et al., 2021; Mhlanga et al., 2019; Muzingili, 2016; Noble, 2016). This reveals an urgent need for action to address this gap. The utilisation of continuous professional development by practitioners can address the disaster-oriented social work knowledge gap. Interestingly, some universities are already making efforts to integrate environmental issues and disaster management in the social work curriculum to equip practitioners for disaster practice. However, some practitioners qualified before this integration, thereby requiring a CPD training programme.

Furthermore, this study also established that the social work profession is not adequately represented in media reports on disaster issues in Zimbabwe. This finding supports the literature asserting that there is limited media representation of social workers' efforts and interventions during disaster times (Hay & Pascoe, 2021; Rasool, 2020). If this continues, the social work profession may not be adequately recognised in disaster management efforts. The practitioners need to break away from their silos and work in collaboration with a multidisciplinary team on disaster management to ensure that their efforts and contributions are recognised and reported, ultimately increasing their media presence.

Additionally, the findings revealed that the social work practitioners are not adequately involved in disaster management efforts. Participants reported that the social work profession is mainly visible and vibrant in the disaster response phase by providing immediate disaster relief such as food and temporary shelter. These findings align with those of Mhlanga et al. (2019), who found that disaster-oriented social work practice in Zimbabwe is concentrated on disaster response, paying little to no attention to disaster prevention and preparedness. This raises important issues regarding the need for continuous professional development among social work practitioners in Zimbabwe for effective disaster practice.

Limitations

This study was insightful; however, there were certain limitations. Firstly, the study had a small sample size, limiting the generalisability of the findings to the entire population of social workers in disaster practice. Additionally, it was challenging to organise the focus group discussions, due to conflicting schedules, resulting in rescheduling to accommodate each participant. Although the study had a few limitations, there were some strengths too. The qualitative depth of the study provided rich insights into the training needs of practitioners.

Recommendations and Conclusion

It has been established that climate change-induced disasters impact the Manicaland Province, as evidenced by the number of disasters that have affected it between 2000 and 2023. However, the practitioners are not adequately prepared for disaster practice, due to the deficit in the social work curriculum. Unless all social work training institutions mainstream climate change and disaster issues, effective disaster social work practice may not be achieved in Zimbabwe. The increasing frequency and intensity of these disasters also make continuous professional development essential in equipping social work practitioners with relevant knowledge and skills to effectively develop proactive strategies for disaster preparedness, prevention, response, and recovery. In line with the organisational learning theory, CPD could strengthen practitioners' capacity to promote sustainability and community resilience in response to climate change-induced disasters.

Furthermore, the study recommends disaster knowledge sharing among social work practitioners, researchers, and educators through platforms such as webinars, seminars, and conferences. Profiling the key social work contributions and successes in the disaster management discourse is critical in ensuring the recognition of the profession. Social workers can also play an important role in influencing disaster policies that protect the well-being of vulnerable communities. Moreover, practitioners should avoid working in silos, since disaster management is complex, thus requiring transdisciplinary collaborations.

The CPD training programme targets social work practitioners who need to advance their knowledge and skills on climate change and disasters. Dziro (2013) acknowledges the importance of CPD in social work, since practitioners practice in an evolving world, making it fundamental for social workers to update their knowledge and skills continuously. The CPD training programme that was developed has seven sessions covering disaster-focused social work aspects, including disaster foundations, vulnerable groups, policy and governance, the disaster management cycle, and learning from past experiences, using case studies of past disaster events.

In addition, the programme is based on a blended learning approach, combining guest lecturers, interactive workshops and discussions, formal classroom lectures and presentations, and field exercises. This ensures that training participants gain practical skills and theoretical knowledge of disaster-focused social work practice. Hence, the practitioners would be able to translate the disaster knowledge and skills into actionable strategies to effectively support communities prior, both during and in the aftermath of disasters.

The Possible Continuous Professional Development Programme

Nyahunda et al. (2019) support breaking barriers that affect social workers' intensive involvement in disaster issues, which often emanate from the absence of disaster components in social work education. As such, responses and recommendations proffered by social work practitioners during data location led to the development of the *Breaking Barriers: Building a Competent Disaster Social Worker* training programme, designed to capacitate practitioners to take up roles in disaster contexts with effectiveness.

The training programme is structured into seven sessions.

Session One—Disaster Conceptualisation, Nature, Causes, Prevalence, and Impact of Disasters

The first session exposes social work practitioners to the literature on disaster concepts to enhance their understanding of natural and its subtypes, as well as manmade disasters. At the end of the session, it is anticipated that the practitioners will be knowledgeable about both the causes and effects of disasters.

Session Two: Concepts in Emergencies, Ethics, Values, and Human Rights

During this session, practitioners are taught foundational concepts in disaster management, ethics, values, and human rights issues. As such, the practitioners are expected to demonstrate an understanding of environmental justice and human rights in emergencies.

Session Three: Disasters and Vulnerable Groups

Disasters' occurrences are increasing at an alarming rate globally, affecting almost every community, but vulnerable populations suffer most. The social work profession works closely with vulnerable groups; thus, it is fundamental for the practitioners to understand vulnerable groups affected by disasters. As a result, the practitioners will be able to develop sound and effective intervention strategies for the different groups, increasing their efficiency in protecting vulnerable populations during emergencies.

Session Four: Disaster Management Theories, Policies, Regional and International Protocols on Disaster Management

This session equips the practitioners with knowledge regarding disaster management policies in Zimbabwe, and existing regional and international protocols on disaster risk reduction. Disaster management theories are also discussed. As a result, the practitioners will be better equipped to contribute to policy change and policy development debates related to disaster management issues.

Session Five: Disaster Management Cycle and Social Workers' Role

Understanding the disaster management cycle is fundamental for social work practitioners. The focus of this session is to enhance the practitioners' understanding of the disaster management cycle and their role in each of the phases.

Session Six: Case Studies and Lessons Learned on the Management of Previous Disasters

In this session, the main aim is to enhance the practitioners' understanding of disaster-oriented social work practice, enabling them to take up disaster management roles confidently. The session may make use of guest lectures, inviting social workers with a track record of success in managing disaster, to share their experiences, achievements, and challenges. Case studies of previous major local, regional, and international disasters will be discussed to understand how they were managed and the roles played by social work professionals and/or social work bodies.

Session Seven: Complexities that Social Workers Encounter in Performing their Work During Disaster Events

This session discusses the challenges encountered in disaster-oriented social work practice. The practitioners are expected to discuss the challenges and provide practical solutions.

Competing Interest

The authors have no conflicts of interest to declare that are relevant to the contents of this chapter.

References

African Development Bank. (2024). *Climate change in Africa*. Retrieved January 6, 2024, from https://www.afdb.org/en/cop25/climate-change-africa

Agritex-NEWU-Ministry of Agriculture of Zimbabwe, USAID/Famine Early Warning System World Food Programme Zimbabwe, & FAO/Sub-Regional Office for Southern and East Africa. (2000). *Assessment of the impact of cyclone eline on the food, agriculture and natural resource sector in Zimbabwe*. Retrieved September 15, 2023, from https://fews.net/sites/default/files/documents/reports/1000050.pdf

Alston, M., Hezeleger, T., & Hargreaves, D. (2019). *Social work and disasters: A handbook for practice. Routledge.*

Bouchard, J., Pretorius, T. B., Kramers-Olen, A. L., Pasmanabhanunni, A., & Stiegler, N. (2023). Global warming and psychotraumatology of natural disasters: The case of the deadly rains and floods of April 2022 in South Africa.

Medical-Psychological Annals, 181(3), 234–239. https://doi.org/10.1016/j.amp.2022.07.004

Braun, V., & Clarke, V. (2006). Using thematic analysis in psychology. *Qualitative Research in Psychology, 3*(2), 77–101. https://doi.org/10.1191/1478088706qp063oa

Braun, V., & Clarke, V. (2012). Thematic analysis. In H. Cooper, P. M. Camic, D. L. Long, A. T. Panter, D. Rindskopf, & K. J. Sher (Eds.), *APA handbook of research methods in psychology, Vol. 2. Research designs: Quantitative, qualitative, neuropsychological, and biological* (pp. 57–71). American Psychological Association. https://doi.org/10.1037/13620-004

Buwerimwe, S. (2023) Cyclone Freddy claims 2 lives in Manicaland. *Newsday*. https://www.newsday.co.zw/local-news/article/200008100/cyclone-freddy-claims-2-lives-in-manicaland

Chatiza, K. (2019). *Cyclone Idai in Zimbabwe: An analysis of policy implications for post-disaster institutional development to strengthen disaster risk management*. Oxfam International.

Dembedza, V. P., Chopera, P., Mapara, J., Mpofu-Hamadziripi, N., Kembo, G., & Macheka, L. (2023). The relationship between climate change induced natural disasters and selected nutrition outcomes: A case of cyclone Idai, Zimbabwe. *BMC Nutrition, 9*(19). https://doi.org/10.1186/s40795-023-00679-z

Dominelli, L. (2011). Climate change: Social workers' roles and contributions to policy debates and interventions. *International Journal of Social Welfare, 20*(4), 430–438. https://doi.org/10.1111/j.1468-2397.2011.00795.x

Dominelli, L. (2014). Learning from our past: Climate change and disaster interventions in practice. In C. Noble, H. Strauss, & B. Littlechild. (Eds.), *Global social work: Crossing borders, blurring boundaries* (pp. 341–351). Sydney University Press.

Dominelli, L. (2016). Greening social work: Linking social and environmental justice in social work theory and practice. In: M. Knutagard, A. Meeuwisse, S. Sunesson & H. Sward (Eds.), *Socialt Arbete: En Grundbok. [Social Work: A Textbook*] (3rd ed., pp. 393–413). Natur och Kultur.

Dominelli, L. (2018). *The Routledge handbook of green social work*. Routledge.

Dominelli, L. (2023). *Social work practice during times of disaster: A transformative green social work model for theory, education and practice in disaster interventions*. Taylor & Francis.

Dziro, C. (2013). Trends in social work education and training: The case of Zimbabwe. *International Journal of Development and Sustainability, 2*, 1423–1435.

Ghazali, D. A., Guericolas, M., Thys, F., Sarasin, F., Arcos González, P., & Casalino, E. (2018). Climate change impacts on disaster and emergency medicine focusing on mitigation disruptive effects: An international perspective. *International Journal of Environmental Research and Public Health, 15*(7), 1379.

Halton, C., Powell, F., & Scanlon, M. (2015). *Continuing professional development in social work*. Policy Press.

Harms, L., & Alston, M. (2018). Post disaster social work. *Australian Social Work, 71*(4), 386–391.

Hay, K., & Pascoe, K. M. (2021). Social workers and disaster management: An Aotearoa New Zealand perspective. *British Journal of Social Work, 51*(5), 1531–1550. https://doi.org/10.1093/bjsw/bcab127

Hay, K., Pascoe, K. M., & McCafferty, L. (2021). Social worker experiences in disaster management: Case studies from Aotearoa New Zealand. *Aotearoa New Zealand Social Work, 33*(1):17–28. https://doi.org/10.11157/anzswj-vol33iss1id820

Huang, Y., Fu, Y., & Wong, H. (2013). Challenges of social workers' involvement in the recovery of the 5.12 Wenchuan Earthquake in China. *International Journal of Social Welfare, 23*(2), 1–11. https://doi.org/10.1111/ijsw.12029

Hudson, C., Goel, K., & Cowie, J. (2021). Making continuing professional development relevant and accessible in regional social work. *Australian Social Work, 74*(1), 29–41. https://doi.org/10.1080/0312407X.2020.1817960

International Federation of Red Cross and Crescent Societies. (2018). *Annual report*. International Federation of Red Cross and Crescent Societies.

International Federation of Red Cross and Red Crescent Societies. (2020). *Zimbabwe: Tropical Cyclone Idai Final Report.* Retrieved August 17, 2023, from https://reliefweb.int/report/zimbabwe/zimbabwe-tropical-cyclone-idai-final-report-dref-operation-n-mdrzw014

International Federation of Red Cross and Red Crescent Societies. (2021). *Final Report Zimbabwe: Tropical Cyclone Eloise.* Retrieved August 17, 2023, from https://reliefweb.int/report/zimbabwe/zimbabwe-tropical-cyclone-eloise-operation-final-report-dref-n-mdrzw015-12-october

Jayawardena, A. W. (2015). Hydro-meteorological disasters: Causes, effects and mitigation measures with special reference to early warning with data driven approaches of forecasting. In *IUTAM Symposium on the Dynamics of Extreme Events Influenced by Climate Change* (pp. 3–12).

Machimbidza, D., Nyahunda, L., & Makhubele, J. (2022). The importance of social work roles in disaster risk management in Zimbabwe. *Technium Social Sciences Journal, 27*(1), 717–726. https://doi.org/10.47577/tssj.v27i1.5554

Maripe, K., & Sinkamba, R. P. (2014). Social work education and disaster risk reduction in Botswana. *International Journal of Multidisciplinary Academic Research, 2*(2), 23–41.

Matlakala, F. K., Nyahunda, L., & Makhubele, J. C. (2021). Challenges faced by social workers dealing with victims and survivors of natural disasters. *Eurasian Journal of Social Sciences, 9*(3), 189–197. https://doi.org/10.15604/ejss.2021.09.03.005

Matlakala, F. K., Makhubele, J. C., & Nyahunda, L. (2022). Social workers' intervention during natural hazards. *Journal of Disaster Risk Studies, 14*(1), 1176.

Matutu, F. (2021). Nature-based solution to disaster risk reduction. Retrieved August 15, 2023, from https://www.preventionweb.net/news/nature-based-solution-disaster-risk-reduction September 18, 2023.

Mavhura, E. (2020). Learning from the tropical cyclones that ravaged Zimbabwe: Policy implications for effective disaster preparedness. *Natural Hazards, 104*, 2261–3227. https://doi.org/10.1007/s11069-020-04271-7

Mavhura, E., & Aryal, K. R. (2022). An adaptation of macroscale methodology to assess the direct economic losses caused by Tropical Cyclone Idai in Zimbabwe. *Journal of Disaster Risk Studies, 14*(1), 1–9. https://doi.org/10.4102/jamba.v14i1.1276

Mhlanga, C., Muzingili, T., & Mpambela, M. (2019). Natural disasters in Zimbabwe: The primer for social work intervention. *African Journal of Social Work, 9*(1), 46–54.

Monash University. (2016). *Social work disaster network (SWDN).* Retrieved January 6, 2024, from https://www.monash.edu/medicine/spahc/social work/archive/news-archive2/news-2016/social-work-disaster-network-swdn

Mukurazhizha, R. M., & Matanga, S. Y. (2023). The role of women in managing the environmental crisis: A case study of Cyclone Idai in Chipinge, Zimbabwe. *Teologiese Studies/Theological Studies, 79*(3), Article a8042. https://doi.org/10.4102/hts.v79i3.8042

Murenje, M. & Porter, S. K. (2020). Covid-19: challenges and prospects for the social work profession. International Federation of Social Workers Press. Retrieved March 21, 2024, from https://www.ifsw.org/wp-content/uploads/2020/04/MurenjePorter_ChallengesAndProspects.pdf.

Muzingili, T. (2016). Greening the profession: A missing perspective in contemporary social work practice in Zimbabwe. *African Journal of Social Work, 6*(2), 29–38.

Noble, C. (2016). Green social work—The next frontier for action. *Social Alternatives, 35*(4), 14–19.

Nyahunda, L., Matlakala, F. K., & Makhubele, J. C. (2019). Role of social workers in mitigating the effects of climate change in Makonde communal lands, Zimbabwe. *Journal of Social Sciences and Humanities, 16*(9), 172–183.

Pyles, L. (2017). Decolonising disaster social work: Environmental justice and community participation. *The British Journal of Social Work, 47*(3), 630–647.

Rambaree, K. (2020). Environmental social work: Implications for accelerating the implementation of sustainable development in social work curricula. *International Journal of Sustainability in Higher Education, 21*(3), 557–574. https://doi.org/10.1108/IJSHE-09-2019-0270

Rapeli, M. (2017). *The role of social work in disaster management in Finland.* Jyväskylä University Printing House.

Rasool, S. (2020). Social workers are an untapped resource to address the psychosocial effects of Covid-19. *Mail & Guardian.* https://mg.co.za/article/2020-04-13-social-workers-are-anuntapped-resource-to-address-the-psychosocial-effects-of-covid-19/

Research in Practice. (2023). *A guide to continuing professional development (CPD) for social workers: Practice guide.* Research in Practice.

Risiro, J. (2019). The challenges of integrating indigenous knowledge in the teaching of weather and climate in geography in Manicaland province of Zimbabwe. *Journal of Geography Education in Africa, 2*, 30–46. https://doi.org/10.46622/jogea.v2i1.2483

Senge, P. (1990). *The fifth discipline: The art and practice of the learning organization.* Doubleday/Currency.

Shokane, A. L. (2019). Social work assessment of climate change: Case of disasters in greater Tzaneen municipality. *Jàmbá: Journal of Disaster Risk Studies, 11*(3), 1–7.

Tevera, D., & Raimundo, I. M. (2021). Struggles of cyclone Idai floods survivors inhabiting spaces of vulnerability and reconstructing their fractured livelihoods. In: G. Nhamo, & D. Chikodzi (Eds.), *Cyclones in Southern Africa. sustainable development goals series.* (pp. 131–142). Springer. https://doi.org/10.1007/978-3-030-74303-1_9

United Nations Office for the Coordination of Humanitarian Affairs (OCHA). (2020). *Southern Africa, Flash Update No.6: Tropical Storm Chalane* (as of 31 December 2020). Retrieved August 15, 2023, from https://reliefweb.int/report/zimbabwe/southern-africa-flash-update-no6-tropical-storm-chalane-31-december-2020

United Nations Office for the Coordination of Humanitarian Affairs (OCHA). (2022). *ZIMBABWE: Floods and Storms Flash Update* No. 1. Retrieved August 15, 2023, from https://www.unocha.org/publications/report/zimbabwe/zimbabwe-floods-and-storms-i-flash-update-no-1-3-february2022#:~:text=The%20Tropical%20Storm%20Ana%20weather,845%20houses%20and%2051%20schools.&text=Tropical%20Storm%20Ana's%20hardest%20hit,Mt%20Darwin%2C%20Mutasa%20and%20Mbire

World Economic Forum. (2024). Quantifying the impact of climate change on human health: Insight report. Retrieved April 22, 2024, from https://www3.weforum.org/docs/WEF_Quantifying_the_Impact_of_Climate_Change_on_Human_Health_2024.pdf

World Health Organisation. (2021). *Mozambique determines priority actions in wake of cyclones to improve future response using after action reviews (AARs)—Case study*. Retrieved October 1, 2023, from https://extranet.who.int/sph/sites/default/files/document-library/document/Mozambique%20determines%20priority%20actions%20in%20wake%20of%20cyclones%20to%20improve%20future%20response%20using%20after%20action%20reviews%20%28AARs%29.pdf

World Meteorological Organization. (2023). *State of climate change in Africa 2022*. Retrieved January 6, 2024, from https://wmo-limbgallery.ovh/records/item/67761-state-of-the-climate-in-africa-2022

World Meteorological Organization. (2024). *Tropical cyclone*. Retrieved January 6, 2024, from https://wmo.int/topics/tropical-cyclone

Zakour, M. (1997). Disaster research in social work. *Journal of Social Service Research, 22*(1–2), 7–25. https://doi.org/10.1300/J079v22n01_02

Zimbabwe National Statistics Agency. (2023). *Zimbabwe 2022 population and housing census report*. Retrieved March 21, 2024, from https://www.zimstat.co.zw/wp-content/uploads/Demography/Census/2022_PHC_Report_27012023_Final.pdf

CHAPTER 6

Centring the Response to the KwaZulu-Natal Floods in the African-Centred Psychosocial Interventions

Sibonsile Zibane, Sthembiso Phoswa, and Nokukhanya Zondi

Introduction

In April 2022, the KwaZulu-Natal (KZN) community in South Africa was devastated by severe flooding and landslides caused by heavy rainfall. The resulting floods wreaked havoc, destroying numerous homes, leaving thousands of people homeless, and causing a significant loss of life. About 448 people died, over 40,000 people were displaced, and over

S. Zibane (✉) · S. Phoswa · N. Zondi
University of KwaZulu-Natal, Durban, South Africa
e-mail: zibanes@ukzn.ac.za

S. Phoswa
e-mail: 215062156@stu.ukzn.ac.za

N. Zondi
e-mail: 213543939@stu.ukzn.ac.za

D. D. Joseph et al. (eds.), *Climate Change*,
https://doi.org/10.1007/978-3-031-83037-2_6

12,000 houses were destroyed (ReliefWeb, 2022). It was one of the most severe flooding incidents experienced in recent history and the deadliest since the floods of 1987. According to the minister in charge of disaster management at the time, Nkosazana Dlamini-Zuma, these floods were the worst floods that had ever been seen in living memory (Grab & Nash, 2023). The KZN flood survivors were left traumatized, displaced, and disoriented, struggling to cope with the aftermath of the catastrophe.

As part of the emergency response and rescue operation, many of the flood survivors were housed in community centres, which, although it provided shelter, also had unintended negative consequences. Being housed in these centres undoubtedly saved the lives of the survivors, providing them with much-needed shelter and protection. However, the circumstances under which they lived within these centres had a profound impact on their well-being.

The centres often lacked the basic necessities and amenities required for a dignified and comfortable existence. For example, the lack of privacy, personal space, sufficient bedding, and proper nutrition in community centres took a toll on the survivors' mental well-being. These conditions compromised the survivors' physical comfort and quality of rest, leading to increased fatigue and vulnerability to health issues. Moreover, the absence of personal space and familiar surroundings, and the loss of familiar social networks added to their sense of displacement and disconnection.

Current trauma-informed theories and interventions in South Africa have been heavily criticized for leaning on Eurocentric epistemologies. The need for a comprehensive and culturally sensitive response to address the distinctive challenges the KZN Flood survivors faced became apparent. This chapter reports on how a team of social workers and psychologists from the Ma'at Institute[1] implemented African-centred Psychosocial Interventions (APIs) for the flood survivors residing in community halls. These interventions were aimed to mitigate and minimize the effects of the floods, facilitate healing, restore a sense of self, and

[1] Inspired by working with vulnerable communities in more than 10 African countries, the Ma'at Institute was established within the School of Applied Human Sciences in the College of Humanities of the University of KwaZulu-Natal, South Africa, to specifically advance Afrocentric thoughts and the provision of African-centred psychosocial services to communities experiencing adversities. The services of Ma'at are multidisciplinary and often involve social work academics, social workers, educational and clinical psychologists, registered counsellors (psychology), and graduate interns.

strengthen the resilience of the flood survivors. Moreover, the APIs were aimed at creating community halls as shared and respectful spaces that promote interactions and activities that facilitate community bonding.

The authors of this chapter are social work researcher-practitioners who were part of the broader disaster response team of the Ma'at Institute. In this chapter, the authors used a collaborative autoethnography to reflect and share the APIs they implemented with African indigenous communities (AICs) who were flood survivors housed in the eThekwini Municipality, KZN province, South Africa community halls. As much as the authors are qualified social workers, they did not view themselves as experts, but they recognized the significance of purposeful learning, affirmation, and documentation of African indigenous knowledge and practices within their interactions with the AICs. Purposeful learning allowed them to acknowledge the value and wisdom embedded in local cultures and traditions, recognizing them as holding valuable insights and solutions that can contribute to addressing the challenges faced by flood survivors.

This chapter addresses several critical questions concerning working with African indigenous communities (AICs) in the context of disaster recovery. First, it explores the methodological approach required to engage effectively with AICs, acknowledging the significance of their unique cultural backgrounds and worldviews. Second, it delves into the process of negotiating entry, establishing relationships, and building rapport within AICs, recognizing the need for trust and collaboration. Lastly, it examines African-centred trauma debriefing and resilience-strengthening tools and techniques that are specifically relevant to AICs.

This chapter suggests moving away from Eurocentric psychosocial services that consider selfhood as distinct from culture and language. Instead, it puts forth a compelling argument for context-specific, holistic, and culturally relevant interventions that uphold and respect the frames of reference, language, and ways of being of the affected individuals. By doing so, the interventions aim to empower the flood survivors, restore their sense of identity, and foster their overall well-being in a manner that aligns with their cultural heritage.

It is against this background that the section below delves into the comprehensive understanding of the factors, influences, and dynamics that shape the need for, and the potential impact of, the APIs discussed in this chapter. By exploring the broader context, we aim to create a solid foundation for readers to comprehend the intricacies and significance of the discussed interventions within its larger framework.

The Broader Context for the Intervention

In the complex fabric of societal well-being, the presence of disasters not only disrupts the equilibrium of normal life but also triggers a series of challenges that affect communities. Bhadra (2021, p. 185) states that, "disaster in human life has various grave impacts that cause threats to the development and normal life of any society". Informed by the International Federation of Social Workers (IFSW) definition of social work, social workers play a vital role in disaster management contexts. They provide psychosocial support, conduct needs assessments, coordinate services, foster community development, mobilize resources, and advocate for the rights and well-being of flood survivors. Their expertise and compassion contribute significantly to the recovery and resilience of individuals and communities impacted by disasters.

The point of departure for any valuable programme is to be centred on the people's lived experiences and culture, inclusively and in a participatory manner. This assertion thus lays the foundation for the main arguments of this chapter, which emphasizes that interventions aimed at African indigenous communities are to be rooted in their culture, traditional practices, ethics, and knowledge (Mkhize et al., 2014). Contextually speaking, this meant that as much as natural disasters are a global phenomenon, the design and the implementation of any response aimed towards African communities should place their interests, history, culture, and experiences at the centre.

To ensure the relevance of interventions to flood survivors, this chapter advocates for the indigenous social work perspectives that are framed within the Afrocentric paradigm. In this chapter we use Afrocentric paradigm, Afrocentric theory, Afrocentrism, and African-centred worldview interchangeably. Afrocentric theory is a framework that emphasizes the centrality of African culture, history, and experiences in understanding and addressing the needs of African people (Asante, 2003). When it comes to the provision of psychosocial services to indigenous African communities (IACs), Afrocentric theory offers valuable insights and approaches that can enhance the effectiveness and cultural relevance of these services.

African-centred psychosocial interventions provide avenues for overcoming the "dualistic partition" or "split between mind and body, nature and culture" and opportunities for embracing openness to dialogue among different epistemic positions (Mbembe, 2016). Indigenous perspectives in social work thus mean that social work knowledge should arise from within the cultures, reflect local traditions and practices, be interpreted within the local frames of reference, and thus be locally relevant and culturally sensitive (Hetherington et al., 2013; Makhanya & Zibane, 2020; Mkhize et al., 2014). Consequently, healing, restoration, and reconstruction become possible by grounding intervention programmes in African ethics, indigenous knowledge systems, principles, and values.

This chapter contends that the idea of indigenizing or "Africanizing" the psychosocial support programme does not simply mean the rejection of Western knowledge or the training we received in our professions, which is predominantly Western. The ideology advanced in this chapter means placing Africa at the centre, historically, culturally, philosophically, and morally (Asante, 1980, 2003; Schiele, 1994).

One of the key principles of Afrocentric theory is the recognition of the interconnectedness between individuals, families, and communities within African societies. It acknowledges that an individual's well-being is intimately tied to their social and cultural context (Ramose, 2002). When providing psychosocial services to indigenous African communities, an Afrocentric approach seeks to understand and address the collective experiences and dynamics of the community. It highlights the importance of incorporating African cultural values, traditions, and belief systems into the therapeutic process. This means recognizing and validating the cultural identity and worldview of the individuals and communities being served. By acknowledging and respecting cultural practices, languages, and ways of being, Afrocentric theory seeks to create a therapeutic environment that is familiar, affirming, and empowering for the clients.

In the context of the KZN floods, the authors of this chapter particularly framed their work and this discussion in the Afrocentric theory because it emphasizes the role of community and social support systems in the healing process. It recognizes that the AICs have their own resilience and collective coping mechanisms, which should be acknowledged and reinforced. Thus, our APIs were not solely focused on individual therapy but also involved engaging with the community, involving community leaders, and promoting community-based interventions that support healing and well-being.

Another important aspect of Afrocentric theory in providing psychosocial services to the KZN Floods survivors is that it allowed for the recognition and understanding of the historical and systemic factors that have shaped the experiences and challenges AICs encounter. This includes addressing the impact of colonization, racism, and other forms of oppression on mental health and well-being. By acknowledging and addressing these broader social and historical contexts, our APIs seek to promote social justice and empower individuals and communities to challenge and overcome systemic barriers.

In the succeeding section, the authors of this chapter share their collective understanding, experience, and approach used in implementing African-centred Psychosocial Interventions (APIs) with African indigenous communities (AICs) who were flood survivors housed in the eThekwini Municipality (EM) community halls.

Methodological Approach

There were 120 community halls which housed flood survivors of EM. It is important to note that the broader interventions by the Ma'at Institute were implemented in both rural and urban areas of EM. However, this chapter reports on interventions in five community halls located in the rural part of EM named Emachobeni, Amatikwe, MzinyathiX2, and Molweni. The number of people housed in each of these halls ranged from 15 to 108 persons per hall. In line with the discussions in this chapter, the five community halls are purposively selected as they met the following criteria:

(a) Participants who were flood survivors and full-time staying at a community hall;
(b) Participants who originally come from the same area, share a similar culture and value system;
(c) Participants who speak the same home language;
(d) Participants who are from marginalized communities, with no access to psychosocial services (In this case, rural communities);
(e) Participants from AICs.

The chapter draws on data generated through collaborative autoethnography (CAE). CAE is best suited to reflect, document, and analyse the interventions by the Ma'at Institute as it emphasizes the importance of honouring and centring Indigenous perspectives, histories, and knowledge in research practices and collaborations (Chang & Nginjiri, 2020). CAE also allowed the authors to act both on behalf of one another and together (active solidarity), to speak to and to listen to each other (communicative solidarity), and to think of each other and with each other (intentional solidarity; Baumann, 2015). Contrary to the dominant Eurocentric epistemologies in social work on how community engagement is conducted, the authors of this chapter adopted Reviere's (2001) five Afrocentric canons to guide their intervention. These are:

Ukweli, which translates to "truth". This canon emphasizes the importance of basing research on the real experiences of the community under study. The process involves recognizing and valuing the lived experiences of community members as authoritative sources that define what is considered true. During the APIs, particularly in the assessment and planning phase, we acknowledged the community members as the primary voices of authority when it came to comprehending how they were impacted by the floods. Instead of relying solely on external perspectives, the experiences and viewpoints of the community members themselves were central. These personal narratives not only shaped the overall story of the impact but also played a crucial role in directing collective efforts to strengthen the community's resilience. In essence, the achievement lies in embracing and prioritizing the genuine experiences of the community, allowing them to guide the research process and subsequent interventions for a more accurate and community-centred approach.

Utulivu is a concept that embodies the principles of justice. It emphasizes the importance of fairness and justice in our APIs. This was achieved because our APIs were culturally relevant to the people served. In dealing with the problems that were out of our scope, we collaborated with other organizations to ensure that a holistic service that views a person as a multidimensional being is provided. For instance, we arranged for medical services to be accessible in the communities we served. This comprehensive approach promoted justice in the provision of services. Additionally, to recognize the significance of language as more than just a means of communication, the use of isiZulu, their indigenous language, was encouraged. As language is seen as a gateway to accessing one's culture, our APIs became the vehicle for acknowledging the importance of cultural preservation and self-expression (wa Thiong'o, 1986). In the spirit of justice, we did not discriminate against our participants as all participants were treated with respect and as active agents.

In line with this understanding, we encouraged the use of isiZulu, the indigenous language of the community, acknowledging language as more than a tool for communication. As argued by wa Thiong'o (1986), language serves as a gateway to culture, making our APIs a platform for embracing cultural preservation and individual self-expression. Through the use of their indigenous language, community members were afforded the agency to articulate their experiences, needs, and aspirations in a manner that resonated authentically with their cultural context. There are isiZulu expressions that lose their meaning if translated to English. For example, the question "uzizwa njani" which forms part of the greetings during APIs, holds a significance that diverges from its English translation of "how are you feeling". "Uzizwa njani" during an API session is "a call for one to pour their soul. In so doing one is activating what is known as *uzwelano. Uzwelano* could be referred to as empathy in Western Psychology but in African ways of knowing it transcends to mean a reciprocal process which allows the listener to get into the pain of the troubled while the troubled benefits from the healing that is activated" (Cakata, 2023, p. 53). Thus, the linguistic inclusivity, served as a form of justice in our APIs' endeavours, ensuring that the voices of the community were not only heard but also valued in a manner that respected their cultural heritage.

Uhaki means harmony. Our APIs were aimed at a holistic approach to care and healing. They acknowledge the interconnectedness of multidimensions of human existence, such as social, cultural, and spiritual aspects. Cakata (2023, p. 51) asserts that "among Africans there is no separation between the physical and the spiritual forms of being". The Western world's separation of body and spirit as captured by Myers (1987) assumes that the various segments of being human could be dealt with individually. APIs assisted our participants in understanding and addressing the multiple facets of individuals and communities in a comprehensive and balanced manner. By incorporating music and dance into our APIs, we recognized that sharing a story about trauma and social needs can leave a person emotionally burdened and vulnerable. Music and dance serve as excellent tools to regulate emotions, alleviate pain, and trigger the release of endogenous opioids in the brain. These elements contribute to a sense of spiritual lightness and help diminish feelings of helplessness.

Fostering Relationships and Empowering IACs Through the Use of Indigenous Language and Ukuhlonipha Language

Establishing professional relationships with traumatized communities is never easy. How introductions are critical to how success you will be in this community. Once the process of '*ukukhuleka*' was concluded and permission to work in the area was granted, we were allowed to introduce ourselves to the flood survivors. We intentionally and purposefully employed isiZulu as the primary language for communication and service provision. Indigenous languages play a vital role in shaping the identity, heritage, and worldview of the IACs. For indigenous peoples, languages not only identify their origin or membership in a community, they also carry ethical values of their ancestors—the indigenous knowledge systems that make them one with the land and are crucial to their survival, hopes, and aspirations (Makhanya & Zibane, 2020).

The use of the isiZulu language creates an atmosphere of mutual respect, enabling meaningful dialogue and establishing a foundation for collaboration. Our competency and fluency in the isiZulu language and Zulu cultural practices were instrumental in breaking the ice, facilitating communication, providing stress relief, and building and enhancing community resilience (Ngcobo et al., 2023, p. 176). By employing language accessible to the participants, we fostered a sense of unity within the community. The indigenous community value collective identity instead of individualistic identity. From the outset of our interactions, we emphasized the use of inclusive concepts like "we" and "our", rather than employing divisive terms like "you", "I", or "they". This approach helped us gain a sense of membership within the community and to display that we subscribe to the African idiom, "I am, because we are".

The experiences of losing their homes and their family members and of being housed in community centres tarnished their dignity and robbed them of their identities. As social workers, we have a shared responsibility to do no harm or revictimization; therefore, we used the '*ukuhlonipha*' language or '*isihloniphiso*' during our interaction with the community. '*Ukuhlonipha*' facilitated a harmonious relationship and strengthened conversation. '*Ukuhlonipha*' is a cultural pillar of African indigenous communities that can be expressed through our cultural linguistic and behavioural aspects of the custom of respect in different spaces (Luthuli, 2007). For instance, we respected the community's values, including being dressed culturally appropriately, sometimes we used '*ukubhinca*' (wrapped in cultural cloth). This came as a surprise to many including social work practitioners, since they were not accustomed to '*ukubhinca*' being a professional dress code in the workplace.

As emphasized by Zondi (2021), when we were engaging with AICs, we did not call adult or authority figures by their first name; rather, their phase of life is followed by their clan names such as '*umkhulu*' (grandfather), '*ukhulu*' (grandmother), '*umama*' (mother), and '*ubaba*' (father), and *mama*. For instance one could be addressed as '*umkhulu* Zibane', while for those in our age group, we would use '*dadewethu (my sister)*' or '*mfowethu*' (my brother). When we were dividing people into small groups for psychological group therapy, we avoided counting people using the index figure as it is perceived as inappropriate behaviour in AICs. In some instances, elders who participated in our APIs commanded the younger participants to observe how we conducted ourselves during our interactions. Showing respect when communicating with indigenous African communities is essential to establish a cordial relationship and trust, which is the essence of negotiating entry and service delivery (Mazibuko, 2008; Zondi, 2021). In the Ma'at Institute, these identities are an expression of respect for the values of the local community.

Building Rapport Through the Use of Izithakazelo (Clan Names)

'Izithakazelo', also referred to as '*izibongo*', '*izithopho*', or '*iziduko*', are praise names or clan names that are traditionally used among various Nguni-speaking ethnic groups in Southern Africa. Mcunu (2004) cited in Mhlambi (2016, p. 111), points out that "izithakazelo are constituted by clan names and by personal praises '*izithopho*'. In performing '*izibongo*', firstly, both '*izithakazelo*' and '*izithopho*' bring about interplay between individual and collective history in matters that concern a group of people who consider themselves a community". In many Nguni cultures, individuals are addressed by their personal name along with their clan's name '*izithakazelo*', which helps establish their identity, ancestry, and social connections (Ndimande-Hlongwa, 2010; Adhikari & Kunene, 2007).

As social workers from AICs, we view our names, surnames, and '*izithakazelo*', as embodying what Mkhize (2018) referred to as being with self and being part of a community. During our APIs, the use of '*izithakazelo*' played a significant role in building and strengthening the relationship with the community members who participated in our services. In every first contact with an individual or a group, one of the first steps was to explore and gather the identifying details of a person. It was at that stage, the moment we knew about the surname of a person, we used '*izithakazelo*'. For example, a person whose surname is Zondi felt "culturally understood" if he or she was referred to in a conversation as Nondaba, Gagashe…which is the clan names of the Zondi clan (Thabede, 2005). Where we were unsure about the '*izithakazelo*' of a surname, we asked the individual to tell us about them.

On many occasions when the individuals were given an opportunity to tell us about their '*izithakazelo*', we observed that the opportunity was an excellent icebreaker. It evoked positive feelings and emotions, which reminded them that despite their current challenges, they are humans, their identities are rooted in their culture and family histories, and the fact that they may feel alone physically but spiritually, they are not alone. Several common experiences and emotions were noted, including a strong sense of pride and connection to one's ancestral lineage and cultural heritage, a sense of belonging and membership within a particular clan and location, memories of their ancestors' contributions, historical events, and legendary figures, as well as an assertion of their identities and place in larger society. It was a common occurrence to hear and see an individual when asked to cite his/her '*izithakazelo*' to use "we" acknowledge that he or she belongs to others. More interesting, we observed that when most people are citing '*izithakazelo*', they cited them through chanting coupled with the traditional movement called '*ukugiya*'. It was such occurrences that offered us opportunities to observe the role of '*izithakazelo*' in the lives of the AICs and how they can possibly build and strengthen resilience.

African-Centred Trauma Debriefing and Resilience-Strengthening Tools and Techniques Relevant to AICs

The significance of employing culturally relevant and context-specific tools and techniques when responding to the needs of AICs cannot be overstated (Zibane, 2017). Such an approach ensures that the assistance or services provided are attuned to the unique cultural norms, values, and circumstances of the affected community. Furthermore, this approach acknowledges and respects the unique cultural, historical, and social contexts of these communities. Thus, employing culturally relevant techniques and tools with individuals who have lost everything because of a disaster becomes the best strategy for trauma debriefing and for strengthening resilience. The following are some of the African-centred tools and techniques that were employed during our African-centred Psychosocial Interventions (APIs) to flood survivors who were housed in the community hall.

Rhythms of Resilience: Drumming Your Way to Strength

During the trauma debriefing sessions with flood survivors, the researcher-practitioners from the Ma'at Institute incorporated drumming or the use of African drums as one of the API techniques. Drumming has long been recognized as a powerful instrument that goes beyond its musical function. In addition to their rhythmic and cultural significance, African writers have extensively explored the transformative and therapeutic qualities of drumming (Summaries, 2019; Ngugi wa Thiong'o, 1986). According to Longdon (2020), researchers have found that if a drum beat frequency of three to four beats per second is sustained for at least fifteen minutes, it will induce significant trance state in most people, releasing our hold on the external world and opening the pathway to receive healing. Moreover, wa Thiong'o (1986, p. 75) asserted that "the drums are the voices of our ancestors, whispering stories of resilience, resistance, and cultural pride".

During the trauma counselling sessions with groups of flood survivors, drumming was used as an icebreaker or/and to induce a meditative state and resilience among the group participants. It was used as a therapeutic approach that harnesses the rhythmic and sensory qualities of drumming to promote physical, emotional, and psychological well-being (Longdon, 2020). Specifically, this technique was incorporated based on the belief that drumming can have various therapeutic benefits, including stress reduction, emotional expression, and the enhancement of overall resilience. As Chinua Achebe, a prominent Nigerian writer, once wrote, the drumbeat carries the heartbeat of our people, reminding us of our inner strength to overcome adversity (Achebe, 2013).

When drumming was introduced, a significant number of our participants seemed to transcend into a spiritual realm. During these moments, clear indications emerged that while physically present within the group, participants were immersed in conversations with unseen entities or their inner selves. When a profound connection with the drums was felt by certain participants, it would prompt them to rise and engage in traditional dance. Others would be inspired to chant, while some would find themselves quietly immersed in introspection and contemplation. Consequently, by the end of the group sessions, most participants will appear relaxed, calm, and less anxious or irritated. Bessie Head, a notable figure from Botswana, eloquently captured the soothing influence of drumming by stating that "the resonance of the drums penetrated the depths of my being, gently easing away the burdens and stresses of the day, enveloping me in a cocoon of serenity" (Head, 2013, p. 108).

Fostering Resilience: The Power of Music and Drama

African music and Zulu dance have long been intrinsic to African cultures, playing a vital role in expressing communal values, rituals, and emotions. Drawing inspiration from the work of Zibane (2017), which delved into the pivotal role of music and rhythm in navigating the challenges of growing in the South African township during apartheid nearly all our APIs with flood survivors incorporated expressive mediums such as music and dance. This was done because African music and dance, with their rich traditions and expressive qualities, serve as powerful tools for fostering resilience and empowering communities (Biko, 1978). Lebaka (2019) also confirmed that song is one of the instruments that African people have utilized to worship and make meaning of their feelings. This assertion is supported by Soyinka (1968) who stated that, African music possesses an "indomitable spirit" that enables individuals and communities to endure and overcome hardships. Through rhythm, melody, and lyrics, African music encapsulates narratives of struggle, triumph, and resilience (Biko, 1978; Zibane, 2017). The use of songs with supportive meanings also played important roles in strengthening the resilience of flood survivors. The lyrics of one of the songs used are as follows:

'Khululeka we Mah,
Khululeka we Mah,
Ungakhali!
Kusekhaya la
Noma ngabe izindlu ezodaka'

English translation is:

(Be free, O Mom,
Be free, O Mom,
Do not cry!
This is home,
Even if we live in mud houses).

The lyrics of this song stirred deep emotions among the participants, evoking a profound sense of hope. Singing and dancing, in addition to fostering a feeling of unity and connection, served as powerful conduits. Throughout our APIs, instances arose where music and dance acted as mediators between anguished flood survivors, our team, and government officials. In these moments, we immersed ourselves in song and dance, channelling our frustrations into rhythmic expressions. With emotions settled, a platform emerged for all parties to voice their concerns and collaboratively seek a resolution. Summaries (2019) illuminated how music and dance intricately bind individuals to their heritage, furnishing them with identity and strength. She underscored that the preservation and embrace of these artistic forms stand as pivotal in nurturing resilience within African communities (ibid). Additionally, Head (2013) acknowledged the transformative potency encapsulated within African music and dance. Of note is the Zulu dance, which she contended to be a conduit for emotional release and healing, guiding individuals through trauma processing and uncovering inner fortitude (Head, 2013).

Voices of Resilience: Unveiling Healing Through Storytelling

From an African-centred perspective, healing is a community endeavour that is achieved through the interdependency of a community, respect, trust, solidarity, caring, uplifting others and the collective memory of the community (Mkhize et al., 2014). Although not unique to AICs, storytelling holds profound significance within African-centred therapy, serving as a powerful and culturally rooted therapeutic tool that facilitates healing, resilience, and personal growth. Embedded within the rich traditions and oral histories of African communities, storytelling goes beyond mere narration; it becomes a vehicle for conveying wisdom, cultural values, and collective experiences (Adichie, 2009).

During the APIs, storytelling was used as a means of allowing flood survivors to share their experiences. The participants sat in a circle form which diminished the power relations between the therapists and the participants, demonstrated equality, and allowed for openness and a free flow of conversation. We also observed that storytelling became an avenue that allowed narrators to engage in self-reflection regarding their personal encounters, while observers acquired valuable perspectives on their own lives by comprehending how individuals have successfully navigated akin adversities.

Conclusion

In the aftermath of the devastating floods and landslides that ravaged the KwaZulu-Natal (KZN) community in April 2022, a poignant narrative of resilience, healing, and restoration emerged through the implementation of African-centred Psychosocial Interventions (APIs). The profound impact of the floods was far-reaching, leaving homes destroyed, lives lost, and a community grappling with trauma, displacement, and a loss of identity. Faced with these immense challenges, a dedicated team of social workers and psychologists from the Ma'at Institute, guided by the tenets of the African-centred worldview, undertook the critical task of tending to the emotional and psychological well-being of flood survivors.

The APIs, grounded in a deep respect for African indigenous communities (AICs), aimed not only to mitigate the immediate effects of the disaster but also to embark on a journey of healing that transcended conventional approaches. Throughout this article, we have explored the multifaceted landscape of working with AICs in the aftermath of a natural catastrophe. The methodological approach highlighted the significance of honouring cultural contexts, language, and traditional ways of being. We delved into the complexities of negotiating entry, building relationships, and fostering rapport within AICs, understanding that trust and understanding are the cornerstones of effective interventions.

In the process, we navigated the terrain of African-centred trauma debriefing and resilience-strengthening tools and techniques, recognizing that true healing lies not in uprooting individuals from their cultural frames of reference, but in creating spaces for them to reclaim their narratives, reclaim their dignity, and reignite their resilience. As this chapter closes, our journey with the flood survivors of KZN resonates as a testament to the power of community, culture, and collective strength. It reaffirms the enduring truth that in the face of adversity, healing is not just a destination, but a transformative odyssey that we undertake together. Our commitment to context-specific, holistic, and culturally relevant interventions remains unwavering, an affirmation that the voices of the KZN community, and countless others like it, will continue to echo through the annals of history as stories of resilience, renewal, and unwavering hope.

References

C Achebe 2013 Arrow of God Penguin

M Adhikari W Kunene 2007 Names and naming in South African history and culture University of KwaZulu-Natal Press

Adichie, C. N. (2009). *The danger of a single story*. TED. https://www.ted.com/talks/chimamanda_adichie_the_danger_of_a_single_story

MK Asante 2003 Afrocentricity: The theory of social change Temple University Press

MK Asante 1980 Intercultural communication: An inquiry into research directions Annals of the International Communication Association 4 1 401 410

G Baumann 2015 Ancient Egyptian Ma'at or old testament deed-consequence nexus as predecessors of ubuntu? Verbum et Ecclesia 36 2 1 4

Bhadra, S. (2021). Disaster management in India: Perspectives from social work practice. In S. Roy (Ed.), *Social work education: The indigenous perspective*. Sage Publication

S Biko 1978 I write what i like Bowerdean Press

Z Cakata 2023 The place of indigenous African languages in the new curriculum: An African psychology case study South African Journal of Higher Education 37 3 43 58

Chang, H., & Ngunjiri, F. W. (Eds.). (2020). *Collaborative autoethnography: Weaving indigenous and multicultural perspectives*. Lexington Books.

Grab, S. W., & Nash, D. J. (2023). A new flood chronology for KwaZulu-Natal (1836–2022): The April 2022 Durban floods in historical context. *South African Geographical Journal*, 1–22.

Hetherington, T., Coates, J., Gray, M., & Bird, M. Y. (2013). *Decolonizing social work*. Ashgate Publishing, Ltd.

B Head 2013 When rain clouds gather Waveland Press

MEK Lebaka 2019 The societal value of art and music in the Bapedi culture and the implications for music in healing as a cultural phenomenon European Journal of Multidisciplinary Studies 4 3 64 76

E Longdon 2020 Vibrational sound healing: Take your sonic vitamins with tuning forks, singing bowls, chakra chants, angelic vibrations, and other sound therapies Healing Art Press

TP Luthuli 2007 Assessing politeness, language and gender in hlonipha (Doctoral dissertation) University of KwaZulu-Natal

T Makhanya S Zibane 2020 Students' voices on how indigenous languages are disfavoured in South African higher education Language Matters 51 1 22 37
GB Mazibuko 2008 Ucwaningo lokuqhathanisa amanoveli omlando ka RRR Dhlomo kanye nezibongo zamakhosi akwazulu ezaqoqwa uNyembezi (1958) (Doctoral dissertation) University of KwaZulu-Natal
A Mbembe 2016 Decolonizing the university: New directions Arts and Humanities in Higher Education 15 1 29 45
Mcunu, T. N. (2004). *The dignity of the human person: A contribution of the theology of Ubuntu to theological anthropology* (Doctoral dissertation, University of South Africa).
IJ Mhlambi 2016 Wena ungubani (Who are you)? Post-1994 identity and memory through ukuthakazela in the 'new' media blog South African Journal of African Languages 36 109 122
N Mkhize S Mathe N Buthelezi 2014 Ethical decision-making in cultural context: Implications for professional practice Mediterranean Journal for Social Sciences 5 23 2413 2419
Mkhize, N. (2018). Ubuntu-botho approach to ethics: An invitation to dialogue. In *African perspectives on ethics for healthcare professionals* (pp. 25–48).
N Ndimande-Hlongwa 2010 Nicknames of South African soccer teams and players as symbols of approbation in a multilingual and multicultural country South African Journal of African Languages 30 1 88 97
Ngcobo, N., Mzinyane, B., & Zibane, S. Z. (2023). *Responding to concurrent disasters: Lessons learnt by social work academics engaging with flood survivors during a COVID-19 pandemic, in South African townships*.
Ngũgĩ wa Thiong'o. (1986). *Decolonising the mind: The politics of language in African literature*. London: James Currey.
Nyembezi, S., & Nxumalo, O. E. H. (1996). *Inqolobane Yesizwe*. Shuter & Shooter.
M Pellerin 2012 Benefits of Afrocentricity in exploring social phenomena: Understanding Afrocentricity as a social science methodology The Journal of Pan African Studies 5 4 149 160
MB Ramose 2002 African philosophy through Ubuntu Mond Books
Reliefweb. (2022). *South Africa: Floods and landslides*—Apr 2022. https://reliefweb.int/disaster/fl-2022-000201-zaf. Retrieved on 05 September 2022.
R Reviere 2001 Toward an Afrocentric research methodology Journal of Black Studies 31 6 709 728
JH Schiele 1994 Afrocentricity as an alternative world view for equality Journal of Progressive Human Services 5 1 5 25
W Soyinka 1968 The writer in an African state Transition 31 11 13
Summaries, B. (2019). *Purple Hibiscus by Chimamanda Ngozi Adichie (Book Analysis): Detailed summary, analysis and reading guide*. BrightSummaries.com.

DG Thabede 2005 Social casework: An Afrocentric perspective (Doctoral dissertation) Stellenbosch University

UNICEF. (2021). Participatory integrated community-based approach development of the approach & process in Niger and Mauritania. https://www.unicef.org/wca/media/8371/file/Participatory%20integrated%20community-based%20approach.pdf. Retrieved on 12 May 2023.

Zibane, S. (2017). *Negotiating sexuality: Informal sexual cultures amongst young people at a township high school in KwaZulu-Natal.* (Doctoral dissertation), University of KwaZulu-Natal, Howard College, South Africa.

Zondi, N. (2021). *Exploring child participation in parental divorce matters in African ind [i] genous communities of KwaZulu-Natal.* (Master's dissertation), University of KwaZulu-Natal, Howard College, South Africa.

PART II

Food Security

CHAPTER 7

Food Security Through Entomophagy: The Benefits and Challenges of Harvesting *Encosternum Delegorguei* in Climate-Change-Affected Rural Nerumedzo Community, Bikita, Zimbabwe

Brenda Nyeverwai Rumutsa

Introduction

Globally, edible insects are an important source of nutritious food and they contribute economic livelihoods (Makore et al., 2023). In Zimbabwe, the need for food is constantly rising due to the burgeoning of population, and a series of hydro-meteorological hazards brought on by climate change has put people in a variety of challenging circumstances and is threatening to make it impossible to fully utilize even the land that is already available (Youdeowei et al., 2000). Ayieko and Oriaro (2008) argue that insects can be part of food security in Africa because they

B. N. Rumutsa (✉)
Parliament of Zimbabwe, Harare, Zimbabwe
e-mail: brendarumucha@yahoo.com

D. D. Joseph et al. (eds.), *Climate Change*,
https://doi.org/10.1007/978-3-031-83037-2_7

contain essential food nutrients such as protein, iron, calcium, riboflavin and zinc which are among the nutrients that are often lacking in the diet of the poor in Africa. This is in consistent with Tchibozo et al. (2005) who revealed that insects can be an important source of animal protein and can be used to improve the health of malnourished children.

In Nerumedzo, Bikita there is also a unique livelihood of collecting the *Encosternum delegorguei*, an edible insect named *Harurwa* in Shona and in Venda *Thongolifa*. This livelihood in Zimbabwe is only peculiar and prominent in Bikita District, specifically in Nerumedzo communal area, as it is the only major collection location in the entire nation. About 85% of the population in Nerumedzo collect the insects not only for household consumption but also for commercial purposes (Makore et al., 2023). The stink bugs are a valuable source of income for the Nerumedzo community and are essential for buying household items and covering school fees (Makuku, 1993). They are also exported to neighbouring countries due to high demand, and collectors are known to travel up to 200 km to areas selling stink bugs (Teffo, 2006). According to Dzerefos et al. (2013) the *Encosternum delegorguei* insects are a high-value food commodity to the *Venda* people of South Africa and earn substantial annual income of up to US$1105 from sales. The price of insects depends on several factors, such as the availability, taste, and labour involved in harvesting.

While previous studies have looked into the nutritional and medicinal value of Harurwa, this chapter investigates how the *Encosternum delegorguei* (Harurwa) harvesting livelihood may effectively enhance food security given the lacuna in the literature that exists on food security through entomophagy. In order to determine the kind of assistance needed to support the livelihood of harvesting insects, it explores the challenges of harvesting Harurwa in an effort to bridge this lacuna.

Climate change and soil exhaustion are the primary global drivers of low agricultural income and food insecurity (Smith et al., 2016). These variables mostly influence those whose livelihoods are dependent on rain-fed agriculture, which is the principal source of income for most people in developing nations, particularly in Africa (Lamptey, 2022). Climate change contributes to poor and uncertain harvests as it increases the frequency of droughts (Serdeczny et al., 2017). It is expected to result in more regular occurrences of extreme dry and hot temperatures, that negatively affect the moisture content of soil (Bharambe et al., 2023).

In Zimbabwe, the majority of rain-fed community farmers are located on less productive arable land (Ndlovu et al., 2020). The decreasing amount of arable land owing to population growth, severe mining, and climate change affecting the country have all negatively impacted the production of croplands (Smith et al., 2016). Temperature extremes, tropical cyclones, storms, floods, and droughts are all cascading from climate change, and as a result, Zimbabwe's rainfall patterns have also been impacted (Ndhovu, 2020). Due to fewer chances for earning a living, impoverished individuals in rural areas typically endure greater hardships than those in urban areas.

Most of the people in the Zimbabwean rural areas are now strengthening their livelihoods through diversifying into marginal activities (Nyahunda et al., 2020). These include selling firewood, brick making, and illegal mining as a way to cover up for the drastic drops of harvests. Such marginal activities have had a serious impact on the environment and also have a retroactive effect on the climate. Most of such non-agricultural livelihoods are now prominent in dry and less productive areas. In some parts of Zimbabwe such as Chivi and Mwenezi people are now relying on the collection of Mopane worms (*Gonimbrasia berlina*). This has managed to strengthen the income and food security for some of the households since the Mopane worms are harvested and sold commercially (Meyer-Rochow et al., 2021).

Purpose of the Study

In view of the existing gap in the literature on food security through entomophagy, this study investigated how the *Encosternum delegorguei* (Harurwa) harvesting livelihood could effectively promote food security. It attempted to fill this gap by examining the benefits and challenges of harvesting Harurwa so as to identifying the type of support required to promote the insect harvesting livelihood.

The Value of Insects

Insects play very fundamental roles in the environment; they deliver a lot of ecological services that are crucial for the survival of humankind. For instance, insects are useful in plant reproduction through pollination and seed dispersal. According to Ingram (1996), estimated pollinator species have been identified and almost all of these (98%) are insects and over 90% of the 250,000 flowering plant species depend on pollinators. Some of the insects serve as sources of food and provide humans with a variety of other valuable products such as silk and medication (Aguilar-Toalá et al. 2022). Additionally vital to agriculture and the natural world is this ecological service provided by insects. Insects have an additional ecological function in the biodegradation of waste. Organic matter is broken down into micro particles by termites, flies, ants, and beetle larvae, which are then ingested by bacteria and fungus (Musbau et al., 2021). By doing so, the minerals and nutrients of dead organisms become readily available in the soil for uptake by plants.

Ayieko and Oriaro (2008) indicate that insects can be part of food security in Africa because they contain essential food nutrients such as protein, iron, calcium, riboflavin, and zinc which are among the nutrients that are often lacking in the diet of the poor in Africa. This is in consistent with Tchibozo et al. (2005) who revealed that insects can be an important source of animal protein and can be used to improve the health of malnourished children. Tchibozo et al. (2005) also adduces that insects provide an excellent source of proteins, minerals, vitamins and energy and their consumption has averted many cases of malnutrition. This is because insects convert plant materials such as leaves, flowers, wood, and others that are not available in palatable or digestible forms, making them available to humans in usable forms whose nutrients are readily available just as is the case with cattle and other large animals. FAO (2010) reveals that while insects have a high protein content comparable to other meat sources, they also contribute nutritious value to diets in traditional and informal food systems. Van Huis (2003) also mentions that insects have played an important role in the history of human nutrition in Africa. According to FAO (2010), an insect's nutritional value varies depending on its species, stage of development, and the kind of foliage that they feed on.

Entomophagy has been used regularly by tribes for centuries as medical treatments and the same practice is being continued in certain countries (Olivadese & Dindo, 2023). The most common insect that has been known for having medicinal purposes is the honey bee. The honey bee sting (venom) relieves arthritis as well as asthma and is also consumed mixed with herbs to treat digestive problems, coughs, colds, sore throats, tuberculosis and lung diseases (Ramos-Elorduy, 1997). Applying honey also on the skin treats excessive scar tissue, burns, and rashes. The bee wax has anti-inflammatory properties and is also used to make antibiotics and anaesthetics (Ramos-Elorduy, 1997). According to Elvin (2005) a rubber-like protein that enables insects to jump, has been used in medicine to repair arteries because of its elastic properties. The other medical applications include maggot therapy and the use of bee products such as honey, propolis, royal jelly, and venom in treating traumatic and infected wounds and burns (van Huis, 2003).

The Chinese people also incorporate herbal use with the use of insects. The centipede in China is used to treat tetanus, seizures, and convulsions (Feng et al., 2000). It is dried and crushed to form a paste that is applied to an afflicted area and is also believed to relieve sores and carbuncles on the skin to alleviate pain especially of headaches and also to cure snake bites. Centipedes are also believed to treat conditions such as ulcers, anaemia, and rheumatic diseases (Ramos-Elorduy, 1988). Termites are also used for indigenous medicinal purposes. The termite mound is dug up and ground into a paste and consumed to treat or cure ulcers, rheumatic diseases, and anaemia (Aja & Padmalatha, 2004). Kumar (2010) reveals that chitin is the most abundant mucopolysaccharide found in insects supporting exoskeletal material. Chitin has a protective effect on candida, an infection caused by yeast, and it also helps in the reduction of plasma cholesterol and triglycerides thus it reduces the intestinal absorption of lipids (Koide, 1998).

According to Van Huis and Oonincx (2017), Insect sales generate higher profits for farmers and, in certain regions, are a significant source of income. For the uneducated and unskilled, harvesting insects offers a source of money and particularly in rural regions, collecting insects provides a remarkable means of employment and revenue generation (Raheem et al., 2019). Raheem (2019) also cites that in many cases insect harvesting can serve as a livelihood diversification strategy that provides both food and income to harvesters.

Sustainable Livelihood Framework

The Sustainable Livelihood Framework serves as the focal point for this study's investigation on the benefits and challenges of venturing into the Harurwa insect harvesting livelihood. The Sustainable Livelihood Framework (SLF) is a conceptual framework developed by the Department for International Development (DFID) of the United Kingdom in the late 1990s. It offers a holistic approach to understanding poverty and development by focusing on the assets and capabilities that people need to sustain their livelihoods. According to Ellis (1998), a sustainable livelihood is one that comprises capabilities, assets, and activities that are required for a means of living. A livelihood is sustainable when it can cope with and recover from stresses and shocks, maintain or enhance its capabilities and assets, while not undermining the natural resource base (Chambers & Conway, 1991).

The Sustainable Livelihood Framework (SLF) brings out the key elements of livelihoods and how they interact (Serrat & Serrat, 2017). These include the assets, strategies, outcomes, policies, and institutions as well as processes. It also identifies entry points for poverty reduction initiatives. Livelihood strategies are activities that people carryout in normal times to make a living such as agriculture and wage labour (Schafer, 2002). Assets encompass what people have, for example, with the Harurwa livelihood this refers to the land which is the harvesting site, the forest product/s which is the Harurwa insect, the physical includes the shelter where harvesters rest in the harvesting season (*Musasa*). Social capital includes the social network where people will sell the insect, the financial asset includes income and savings from selling the insects, and human assets include skills and health. Policies, institutions, and processes determine the access and control of assets and they can contribute to increase in vulnerability or resilience. According to the Sustainable Livelihood Framework (SLF) a livelihood also includes access to, and benefits derived from, social and public services provided by the state such as education, health services, roads, and water supplies (Lipton & van der Gaag, 1993).

Harvesting of Harurwa in Nerumedzo Village

Encosternum delegorguei (*Harurwa*) insects are collected from the wild in the morning and evening in Nerumedzo communal area. This is the time when the temperature is low and the insects are less active making it simple to capture them. Men and children are usually the ones who collect these insects in the area. When the Harurwa insects reach maturity, they can be harvested from April through September. This converges with Selaledi's et al. (2021) findings, which indicate that Harurwa insects arrive between April and September and then migrate southward as the rainy season approaches. The insects according to the Nerumedzo community arrive in a dense mass already mature from unknown breeding grounds. The forests in the Nerumedzo village of Bikita District, where insects are gathered, are regarded as sacred (Makuku, 1993). This culture promotes sustainable use of the natural resource base and it aligns with the sustainable livelihoods framework (SLF) that enunciates that a livelihood should not undermine the natural resource base.

The Research Methodology

This cross-sectional study was conducted between 2022 and 2023 in Zimbabwe's Nerumedzo communal area, which is located in Bikita District. The key research questions were: what are the benefits and challenges of harvesting *Harurwa* insects in the climate-change-affected Nerumedzo communal area? The researcher used a hybrid approach applying both qualitative and quantitative research methods. In-depth interviews and non-participant observations were the qualitative approaches utilized to gather data, and a survey questionnaire was used to obtain quantitative data. For qualitative data, purposive sampling was employed, while random sampling was used for quantitative data. Using a list of harvesters registered at the site, the researcher randomly picked 40 households for the sample. Each home had an equal chance of being chosen and a 30% presentation that Zikmund (2003) and Creswell (2003) recommend was attained. The selection of households was based on the fact that most harvesting and consumption is done at the household level. Thematic analysis was used to analyse qualitative data, and Microsoft Excel was used to analyse quantitative data.

Informed Consent

Informed written consent was solicited from Nerumedzo community leaders. The researcher also sought consent from individual interviewees and participants before data collection. Participants and interviewees were respected and urged to participate on a voluntary basis. If they had some reservations, they were excused. Privacy and confidentiality were maintained throughout the study.

Demographic Profile of the Participants

Out of the 40 sampled respondents, 82% were males and 18% were female recipients. This can be explained by the Zimbabwean culture that males are the ones who are supposed to fend for their families (Shire, 2016). In addition, the African custom in Zimbabwe dictates that only women on menopause are permitted to gather insects at the Nerumedzo *Harurwa* gathering site. Women typically stay behind to cook and take care of other household chores in support of the men who will be harvesting, especially their husbands, as they are not permitted to prepare anything else other than Harurwa in the harvesting site.

Although women must assist men at work, this reinforces the idea that men should work and women should stay at home so they can care for their children. In this instance, the majority of those harvesting were middle-aged, making up 58% of the studied population and falling between the ages of 41 and 59 years. Potential chances in the city tend to pull the young, physically fit people between the ages of 20 and 40 years, leaving the elderly behind. Only a small percentage of people over were above the age of 60 and most of them were pensioners. Since rural areas in Africa offer cheaper living, the majority of retirees decide to stay there after they retire (Greiner, 2011). All of the participants in this study went through primary school level. In Zimbabwe after independence the government made free primary level compulsory and this led to a high literacy rate in the country. Nerumedzo community was not left out on free primary education for all to ensure that everyone could read and write.

Discussion of Results

The section is organized based on the topics that surfaced from thematic analysis and is also informed by the SLF's expected livelihood outcomes. The listed livelihood outcomes on the framework include sustainable use of the natural resource base (Environment), income, well-being, vulnerable reduction, and food security.

Benefits to the Natural Resource Base (Environment)

One of the benefits noted in this study is the preservation of the Nerumedzo forest in which the harvesting is done. A sustainable livelihood according to the SLF should protect the natural resource base on which it depends on, rather than endanger it (Chambers & Conway, 1991). Contrarily, if the natural state of the environment is altered, the Harurwa insects will be forced to relocate and change their habitat in search of another suitable environment. The livelihood will be disrupted if the insects move and this compromises food security. The Nerumedzo community is harvesting the insects sustainably so as to prevent them shifting from their natural habitat.

Another interesting benefit of harvesting the insects highlighted by the participants is that the waste water from killing the Harurwa insects can be used as an insecticide as it deters aphids from leaf vegetables. 35% of the respondents highlighted that the waste water from killing Harurwa during harvesting can be used as an insecticide. This could be explained by the defence odour secreted by the insects when they want to deter predators. One participant specifically said that:

> Aphids are deterred if you use the water used to kill Harurwa to your leafy vegetables, particularly chomolia. Because of this we have actually reduced the use of insecticides.

According to Mazid et al., (2011) generally synthetic pesticides are being extensively used though they are regarded as ecologically unacceptable. Flora and fauna have been adversely affected by the use of synthetic pesticides (Pimentel et al., 1997) and now there is ecological protection pressure to reduce their usage (Ulber et al., 2010). Increasing public concern about the potential adverse effects of synthetic agrochemicals stimulates search for the technologies and such products safer for the environment. The use of pesticides in the area has also been reduced as the harvesters are relying on the biopesticides from the Harurwa secretion. According to Zacharia (2011) some keystone species have been lost due to heavy reliance on some synthetic pesticides. The connections held by keystone species maintain the organization and structure of communities and their loss will result in a range of dramatic cascading effects that alter trophic dynamics and other food web connections and can cause extinction of many species in an ecosystem (ibid). Due to the intensive release of synthetic pesticides into the environment, some of them end up being widely distributed and are found around the globe because of their long-range transboundary atmospheric mobility (Sharma et al. 2020).

The Harurwa biopesticide requires no money to purchase, is an eco-friendly alternative and the people's means of survival is protected as the environment is not jeopardized. This is inconsistent with Chambers and Conway (1991) who emphasize that a livelihood should be sustainable while not undermining the natural resource base. The primary advantage of using the biopesticides from the Harurwa secretion is reduced impact on the environment compared to synthetic chemical pesticides. According to Essiedu et al. (2020) biopesticides are typically derived from natural sources such as plants, bacteria, or fungi, making them biodegradable and less harmful to non-target organisms. By relying on biopesticides, farmers can minimize the contamination of soil, water, and air with toxic chemicals, thereby promoting a more sustainable agricultural system. Another crucial aspect of transitioning to biopesticides is the potential health benefits for the Nerumedzo communities and their surroundings. Conventional pesticides often contain harmful ingredients that can pose risks to human health through exposure during application, consumption of contaminated produce, or environmental pollution (Sharma et al., 2020). In contrast, biopesticides are generally considered safer for use due to their lower toxicity levels and reduced residual effects (Lengai & Muthomi, 2018). Despite being derived from natural sources, biopesticides have demonstrated effectiveness in controlling pests and diseases in agricultural settings. The compounds present in the Harurwa secretion may possess insecticidal, fungicidal, or bactericidal properties that target specific pests while minimizing harm to beneficial organisms. Through proper application and integration into pest management strategies, biopesticides can offer sustainable solutions for pest control without compromising crop yield or quality.

Income (Economic) Benefits

It was noted in this study that as the household size increases, more effort is required to feed the family. This gives a family pressure to search for available livelihood alternatives to secure food. This is in line with the sustainable livelihood framework (SLF), which emphasizes the need for people to be able to identify the points of entry for initiatives aimed at alleviating poverty. People are able to safeguard their sustenance and means of survival once they recognize these alternatives. Most of the people harvested the insects for cash income and consumption. 50% respondents harvested the Harurwa insects for cash income while 45% harvested for consumption. Only 5% of the respondents harvested the insects for barter trade. The findings indicate that 70% of harvesters sell their insects at the local centre and 22% in urban areas; the harvesters do not sell their insects in the neighbouring villages. The least (7%) of the respondents indicated that they sold their insects out of Zimbabwe. The respondents reported that the stated area outside of the nation where they travel to sell the insects is South Africa, to the Venda people in the province of Limpopo.

The harvesters indicated that the mean number of buckets gathered per month was 9. They quantified that the insects are sold for a minimum of $50 for a 25 *l* bucket and the price also depends on where they will be selling. As per results shown one respondent got an average of less than a thousand in a harvesting season for Harurwa. According to the research, the majority of those who harvest Harurwa insects often make between $1000 and $1200 per harvesting season. This is in consistent with the results from a study that was carried out in South Africa by Dzerefos et al. (2013) who reveal that harvesters of the same insects in South Africa earn substantial annual income of up to US$1105 from sales. Two harvesters from the Nerumedzo community who made more money than the others, indicated that they were able to obtain an exceptional amount of $1500. Dzerefos (2019) claims that in many cases insect harvesting can serve as a livelihood diversification strategy that provides both food and income to harvesters. The Harurwa insects are a high-value food commodity and they provide valuable buffers against seasonal shortages of food (Dufour & Paoletti, 2002). The highest number of respondents (25 out of 40) indicated that they use the income that they get from selling the insects to purchase food thus promoting food security and reducing their vulnerabilities. These are followed by 13 respondents who indicated that they use the income to purchase agriculture implements. 6 respondents said they use it to buy clothing while 3 respondents only, highlighted that they use the proceeds from the sale of the insects to pay school fees, medical services and purchase prescription drugs. According to the Sustainable Livelihood Framework (SLF) a livelihood should also include access to amenities such as health services (Lipton & van der Gaag, 1993).

The following Fig. 7.1 is a graph showing the major uses of cash obtained from harvesting insects in Nerumedzo.

These outcomes correspond with findings from the literature. For example, Van Huis and Oonincx (2017) observed that sales of insects increase farmer income and, in some areas, constitute a substantial source of revenue. Insect harvesting is a lucrative activity for the illiterate and unskilled, and it offers a great source of employment and income generation, especially in rural areas (Raheem et al., 2019). According to Raheem (2019), gathering insects can frequently be used to diversify one's income

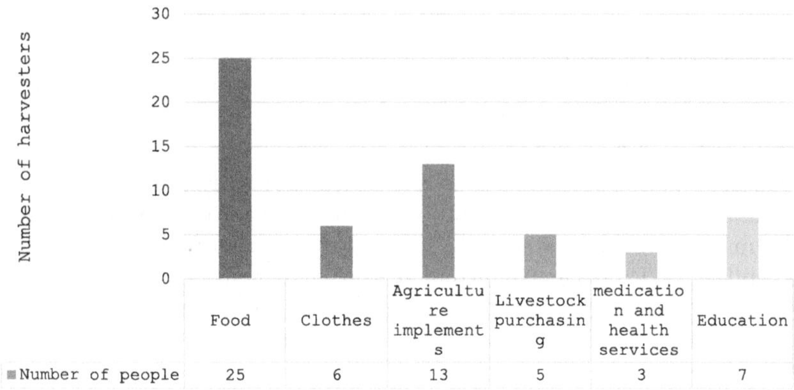

Fig. 7.1 The major uses of cash obtained from harvesting insects in Nerumedzo-Brenda Nyeverwai Rumutsa

and means of subsistence, giving harvesters access to both food and income.

Well-Being

The Oxford English Dictionary defines well-being as "the state of being comfortable, healthy, or happy", however this definition is much broader as it encompasses other aspects as well, such as people's feeling of purpose, overall life satisfaction, and sense of control. Income improves people's well-being and makes them less susceptible to food insecurity. Besides income benefits, 12.5% of the participants indicated that they consumed *Harurwa* insects for medicinal reasons. The respondents listed several medical conditions for which the insect is said to be beneficial, including blood pressure reduction, asthma relief, chest pain relief, and protection against diarrhoeal illnesses. This could be true since the result aligns with literature findings. Olivadese and Dindo (2023), for example, mentioned that entomophagy has been a common medical technique utilized by tribes for ages, and that it is still used in some nations. Other respondents also believed that eating the insects improves appetite. Some people also favoured the insects to other relish or snack. The bitter taste of the insects was preferred especially by those who drink beer. 15% per cent of the respondents also said that harvesting periods allow individuals to socialize and exchange life advice and ideas, which improves social life and additionally helps the lonely especially those that would have retired and the unemployed. Furthermore, with the use of sustainable agricultural practices, such as the application of the previously described biopesticides obtained from the Harurwa secretion, the Nerumedzo population can also enhance and promote their well-being.

Vulnerability Reduction and Food Security

The results also show that out of the 40 respondents 57.5% consumed the Harurwa insects simply because they were in season and 20% people preferred the insects to other relish. 5% people also mentioned that they consumed them for cultural and traditional purposes as it is considered food provision by the ancestors of that area. The research showed that there is income acquired from selling the insects as well as other benefits of harvesting the Harurwa in the Nerumedzo areas. The findings further indicated that 60% of the participants could barter trade the insects for other food items as a way to mitigate against hunger. Since it is becoming difficult to depend only on rain-fed crop production as espoused by Ndlovu et al. (2020). Harvesting Harurwa insects is an excellent opportunity for people who are unemployed, have low-paying jobs, or are retired to supplement their income and food. The advantage of harvesting Harurwa is that it requires trivial livelihood capital assets, e.g. physical and financial assets hence the poor are also able to venture into the livelihood. People who engage in this activity will benefit financially from it; this, in turn, reduces their vulnerability to food insecurity.

Challenges Encountered in Harvesting and Marketing Harurwa Insects in Nerumedzo Community

There are also challenges associated with Harurwa insect livelihood apart from the benefits. The following section presents the challenges identified from the data gathered from the study site.

Challenges Associated with Harvesting

The Harurwa insects have a defence mechanism of stinking; they expel a foul-smelling compound from special thoracic glands when threatened and this is meant to deter predators. According to Taylor (1992) this smell sends a chemical message to other stink bugs, alerting them that there is danger. These scent glands have a major role in reproduction and they play a role in attracting mates. The harvesters of stink bugs indicated that during harvesting the foul-smelling compound stains their hands. The hands become unappealing and the pigment takes days to remove. In relation to this one participant said:

> Being well-fed and having bleached hands is preferable to starving and preserving your hands.

Even though they end up with bleached hands they are willing to continue harvesting as the benefits outweigh the negative consequences. Fortunately, the stains are not permanent and they disappear after the harvesting season.

Harvesters climb trees to harvest in an environmentally responsible manner. Since this is done in the morning, the insects fall to the ground when they shake the branches. To keep the insects from taking off, they are collected and put in buckets of water. Unfortunately, reports of people falling from trees in an attempt to harvest the insects have been reported. Since the insects are only inactive in low temperatures thus during the day and in the evening the harvesters also have a challenge in that their harvesting time is minimal.

Challenges Associated with Marketing

An enquiry was also made to identify the problems faced by the people in Nerumedzo community to market the Harurwa insects. One of the problems identified in marketing Harurwa, is the people's preference for modern foods as opposed to the traditional food such as the insects. This has dwindled the market for selling the insects. The primary marketing problems are that the market for insects is shrinking due to people's preference for modern foods over traditional dishes and because insects are sold in an open market outside of stores, which makes most customers avoid them. According to respondent 14,

> Due to its seeming unpalatability to those accustomed to flavourful food, people no longer favour such traditional foods, particularly Harurwa.

Most people no longer prefer such food, this is in line with Bisconsin-Júnior et al. (2022), who attests that the consumption of insects (entomophagy) is disappearing in some areas as a result of shifting social structures, and rising consumption of contemporary foods. Since the insects are also sold in open markets, most people shun them and regard them as not prepared under proper hygiene conditions. One respondent iterated:

> Sellers of Harurwa do not cover their products with nets, and they measure the quantity customers need using unclean containers. I don't think the containers are regularly cleaned.

Hoffman (2019) is of the idea that food hygiene is indeed a crucial aspect of food production, handling, and marketing, as it directly impacts consumer perception and the desirability of food products. Outside markets are usually a source of disease outbreaks such as cholera and typhoid; therefore, some people do not want to eat foods from outside markets. Instead, they prefer food from shops and restaurants.

Another challenge is that some people from other areas are not familiar with the Bikita delicacy; unfortunately, people do not usually want to buy what they do not know. The respondents also stated that the *Encosternum delegorguei* insects are peculiar and prominent only in Bikita District, one respondent said:

> This delicacy is only found in abundance in Bikita, and only in Bikita and that makes this place special.

Certain environmental conditions or vegetation types in Bikita District create a favourable habitat for Harurwa, leading to their prominence in the area. As highlighted by Makuku (1993), Bikita District, experiences a climate that is conducive to the survival and reproduction of Harurwa. This may include warm temperatures and moderate rainfall, which support the growth of vegetation that these insects feed on.

Finally, since the marking of the beginning of the Harurwa insect harvesting season in the Nerumedzo area involves carrying out a cultural ritual and brewing beer, some Christian groups such as the Apostles and Zionists do not eat the insects because of the spiritual and religious reasons. The other challenge in marketing the insects in Nerumedzo community is that of poor road and communication infrastructure. The road to the harvesting site is slippery and inaccessible especially when it is rainy.

Discussion and Conclusion

The key findings of this study are that the benefits of harvesting the Harurwa insects outweigh the challenges. This supports the claims made by Ayieko and Oriaro (2008) and Tchibodzo et al. (2005) that insects can contribute to food security in Africa since they are a good source of critical nutrients that are frequently absent from the diets of the continent's impoverished. The significant benefit is that the insect acts as a safety net for food security. The Nerumedzo community has a commensurate relationship with the environment, as they are making every effort to harvest insects sustainably in order to avoid disrupting the livelihood as it will compromise food security in the area. As they have recognized the benefits associated with the Harurwa insect livelihood, they have become conscious of the importance of protecting the environment. The harvesting of insects requires policy support and improvement in the management of the natural resource base (Ayeiko, 2014). Policymakers need to develop clear and comprehensive regulatory frameworks that govern the harvesting, processing, and marketing of insects. This includes regulations related to food safety, labelling, certification, and quality standards to ensure consumer protection and public health.

There is a need for greater investment in incorporating DRR into long-term sustainable development so as to cushion the existing livelihoods to reduce vulnerabilities. If the natural resource base is protected from destruction and unsustainable activities, the livelihood is preserved. By safeguarding the natural resource base, such as the habitats where insects are harvested or farmed, communities can ensure that these resources remain available for future generations. Most of the participants highlighted that they mainly harvest the insects for cash income and consumption. They also highlighted that the harvesting periods allow individuals to socialize and exchange life advice and ideas, which improves social life and additionally helps the lonely especially those that would have retired and the unemployed. Several medical conditions for which the insect is said to be beneficial were listed and these include blood pressure reduction, asthma relief, chest pain relief, and protection against diarrhoeal illnesses.

This is congruent with the findings of Muguti (2023), who reports that some individuals eat the insects for the same purpose.

Despite these benefits the challenges highlighted are that the insect secretes a foul deterrent that bleaches their hands and in long term results in the removal of nails. There has also been some reports of people falling from trees trying to harvest the insects. In terms of marketing the main challenges are that people now prefer modern foods much more than traditional food dwindling the market of the insect and the insects are sold in an open market outside shops; therefore, most people shun them. This corresponds with findings by Bisconsin-Júnior et al. (2022) who mentions that the practice of eating insects (entomophagy) is becoming uncommon in some regions due to increasing adoption of modern foods, changed social structures, and changes in demography. Western societies believe that eating insects is a survival strategy used by African primitive people who are struggling to survive (Ayeiko, 2014). Some people in Africa consume insects due to the problems of relish especially in the rustic rural areas due to poverty especially in the areas where vegetables do not grow well and where they cannot afford to slaughter their livestock on a regular basis.

They also regard them as not prepared under proper hygiene conditions. Another challenge is that there is a poor road and communication network to the harvesting site. Although entomophagy is shunned by many people, evidently it can play an important function in addressing food insecurity. It is important to promote sustainable and environmental sensitive livelihood complementary strategies considering the climate change risk drivers that are jeopardizing food and income security of most of rural households. According to Kenney et al. (2023) addressing these climate change risk drivers requires a comprehensive approach that integrates climate adaptation and mitigation strategies, promotes sustainable agricultural practices, strengthens social safety nets, enhances access to financial services and technology, and fosters international cooperation to build resilience and protect the livelihoods of rural households. Food insecurity is a critical issue for most people residing in the rural areas where food production has been lowered by negative effects of climate change, population expansion, and the decline in crop productivity due to reduced soil fertility. Edible insects are an undermined source of livelihood but offers food and income.

Recommendations

Due to their high nutritional value and marketability, insects can greatly improve the standard of living for some rural families. The results of this empirical study showed that harvesting *Encosternum delegorguei* insects can significantly reduce vulnerabilities while strengthening the livelihoods that could be relying on rain-fed agricultural production which has been impacted by climate change. Social workers can facilitate community engagement and participation in insect harvesting initiatives, ensuring that local voices are heard, and communities have ownership over decision-making processes. This involves collaborating with community members, conducting needs assessments, and promoting inclusive approaches to development.

There is a need for the Governments of developing nations to include edible insects in their agenda on addressing food insecurity. These have to promote research, involve relevant stakeholders, and formulate legislation that protects the livelihoods. This disaster risk reduction strategy can go a long way in providing a backup in the case of a hydro-meteorological disaster affecting the area, as people will not be forced to sell their assets in trying to cope with the disasters. Governments, donors, foundations, and migrant groups must collaborate to achieve social transformation in insect harvesting communities. There is a wide-ranging scope within the parameters of existing rural development policies to support, promote, and address the identified challenges that the Harurwa insect harvesters are facing in Nerumedzo. Such actions include increasing the role of the state in particular economic decisions. The Governments must build a strong infrastructure (roads, power, and communications) base to support the mobility and marketing of the insects. In the Nerumedzo community scenario, the government can work with private sectors to establish an infrastructure that will allow easy marketing, as well as training in processing standards and packaging the product so that it can also be sold formally. The production of biopesticides should be promoted as these can be produced at a cost lower than the existing synthetic insecticides, and they are likely to be adopted by the large numbers of poor farmers. Researchers should carry out experiments to explore the efficacy of the Harurwa insect secretion as it could be a potential commercial cheap eco-friendly supplement or an alternative of some synthetic pesticides in crop protection that can promote sustainable agricultural practices.

Further Research

Further research on the medicinal properties of *Encosternum delegorguei* insects is necessary, as harvesters believe that the insects can alleviate and heal specific ailments. Furthermore, as the insects are harvested during winter, additional research on their life cycle is required to determine whether their abundance is affected by climate change and whether rising temperatures negatively impact their proliferation.

Acknowledgements I would like to thank Chief Nerumedzo and the community leaders for allowing me to collect the Encosternum delegorguei insect samples as well as giving me the information that was needed for my research. I thank Mr. Ernest Gwekwe for his technical assistance, my supervisor Mr Paul Chipangura for his patience, insightful comments, and advice. To my husband Maxwell Rumutsa, thank you for empowering me by sending me to the University to pursue my studies, thank you once again for allowing me such a chance. My brothers-in-law Tafadzwa and Silas, thank you for taking care of the children especially Junior (Maxwell) when I was busy with my book chapter, I know it was not easy.

References

JE Aguilar-Toalá RG Cruz-Monterrosa AM Liceaga 2022 Beyond human nutrition of edible insects: Health benefits and safety aspects Insects 13 11 1007

Aja, R. S., & Padmalatha, C. (2004). *Ethno-entomological practices in Tirunelveli district, Tamil Nadu.*

Ayieko, A. (2014). *Review on community food security and edible insects' resources: Biodiversity and policy implications for safeguarding human consumption.*

MA Ayieko V Oriaro 2008 Consumption, indigenous knowledge and cultural values of the lake fly species within the lake Victoria region African Journal of Environmental Science and Technology 2 282 286

KP Bharambe Y Shimizu SA Kantoush T Sumi M Saber 2023 Impacts of climate change on drought and its consequences on the agricultural crop under worst-case scenario over the Godavari River Basin India. Climate Services 32 100415

A Bisconsin-Júnior H Rodrigues JH Behrens MAAP Silva da LRB Mariutti 2022 Food made with edible insects: Exploring the social representation of entomophagy where it is unfamiliar Appetite 173 106001

Chambers, R., & Conway, G. (1991). *Sustainable livelihood-chambers and conway1991. pdf.*

JW Creswell 2003 Research design: Qualitative and quantitative and mixed methods Sage publications

Dufour, D. L., & Paoletti, M. G. (2002). Minilivestock. In D. Pimental (Eds.), *Encyclopedia of Pest Management* (Vol. 1, pp. 487–492). Marcel Dekker.

CM Dzerefos ETF Witkowski R Toms 2013 Comparative ethnoentomology of edible stinkbugs in southern Africa and sustainable management considerations Journal of Ethnobiology and Ethnomedicine 9 1 1 12

F Ellis 1998 Household strategies and rural livelihood diversification The Journal of Development Studies 35 1 1 38

KE Elvin 2005 Nutritional potentialities of the larvae of *Rhynchophorus phoenicis* (F.) Pakistan Journal of Nutrition 4 287 290

Essiedu, J. A., Adepoju, F. O., & Ivantsova, M. N. (2020). Benefits and limitations in using biopesticides: A review. In *AIP conference proceedings* (Vol. 2313, No. 1). AIP Publishing.

FAO (2010). In Forest insects as food; Humans bite back, In *Proceedings of a workshop on Asia pacific resources and their potential for development.* 19–21 February 2008.

Y Feng X Chen S Wang S Ye Y Chen 2000 The common edible insects of Hemiptera and their nutritive value Forest Research 13 608 612

C Greiner 2011 Migration, translocal networks and socio-economic stratification in Namibia Africa 81 4 606 627

CB Ingram 1996 Putting insects on the Australian menu Food Australia 41 565 566

Kenney, C., Phibbs, S., Meo-Sewabu, L., Awatere, S., McCarthy, M., Kaiser, L., Harmsworth, G., Harcourt, N., Taylor, L., Douglas, N., & Kereopa, L. (2023). Indigenous approaches to disaster risk reduction, community sustainability, and climate change resilience. In *Disaster risk reduction for resilience: Climate change and disaster risk adaptation* (pp. 37–59). Springer International Publishing.

Koide, S. S. (1998). Chitin-chitosan: Properties, benefits and risks. *Nutrition Research, 18*

P Kumar 2010 Functioning of public distribution system in India: An empirical evaluation Outlook on Agriculture 39 177 184

S Lamptey 2022 Agronomic practices in soil water management for sustainable crop production under rain fed agriculture of Drylands in Sub-Sahara Africa African Journal of Agricultural Research 18 1 18 26

GM Lengai JW Muthomi 2018 Biopesticides and their role in sustainable agricultural production Journal of Biosciences and Medicines 6 06 7

Lipton, M., & van der Gaag, J. (Eds.). (1993). *Including the poor: Proceedings of a symposium organized by the World Bank and the International Food Policy Research Institute*. World Bank Publications.

Makuku, S. J. (1993). All this for a bug! Community approaches to common property resources management: The case of the Nerumedzo community in Bikita, Zimbabwe. *Forests, Trees and People Newsletter*, *22*.

M Masocha T Dube T Maziva 2018 Integrating environmental variables and geospatial technologies in landscape scale habitat modelling of edible stink bugs in Zimbabwe Physics and Chemistry of the Earth, Parts A/B/C 105 206 211

M Mazid TA Khan F Mohammad 2011 Role of secondary metabolites in defense mechanisms of plants Biology and Medicine 3 2 232 249

VB Meyer-Rochow RT Gahukar S Ghosh C Jung 2021 Chemical composition, nutrient quality and acceptability of edible insects are affected by species, developmental stage, gender, diet, and processing method Foods 10 5 1036

SA Musbau BH Ayinde OO Omowunmi EL Motunrayo E Belay B Teshome MT Sarwar 2021 Micro and macro (organisms) and their contributions to soil fertility Front Environ Microbiol 7 44

Ndlovu, E., Prinsloo, B., & Le Roux, T. (2020). Impact of climate change and variability on traditional farming systems: Farmers' perceptions from south-west, semi-arid Zimbabwe. *Jàmbá: Journal of Disaster Risk Studies*, *12*(1), 1–19.

C Nhemachena L Nhamo G Matchaya CR Nhemachena B Muchara ST Karuaihe S Mpandeli 2020 Climate change impacts on water and agriculture sectors in Southern Africa: Threats and opportunities for sustainable development Water 12 10 2673

Nyahunda, L., Makhubele, J. C., FK, M., & Mabvurira, V. (2020). Resilience strategies of rural people in the face of climate change in Mazungunye community, Ward 4, Bikita District, Masvingo Province Zimbabwe: A social work perspective. *Gender and Behaviour*, *18*(2), 15511–15520.

M Olivadese ML Dindo 2023 Edible insects: A Historical and cultural perspective on entomophagy with a focus on Western societies Insects 14 8 690

Pimentel, D., Wilson, C., McCullum, C., Huang, R., Dwen, P., Flack, J., Tran, Q., Saltman, T., & Cliff, B. (1997). Economic and environmental benefits of biodiversity. *BioScience*, *47*(11), 747–757.

D Raheem C Carrascosa OB Oluwole M Nieuwland A Saraiva R Millán A Raposo 2019 Traditional consumption of and rearing edible insects in Africa, Asia and Europe Critical Reviews in Food Science and Nutrition 59 14 2169 2188

J Ramos-Elorduy 1997 Importance of edible insects in the nutrition and economy of people of the rural areas of Mexico Ecology of Food and Nutrition 36 347 366

Schafer, J. (2002). *Supporting livelihoods in situations of chronic conflict and political instability: Overview of conceptual issues*. Overseas Development Inst.

L Selaledi Z Hassan TG Manyelo M Mabelebele 2021 Insects' production, consumption, policy, and sustainability: What have we learned from the indigenous knowledge systems? Insects 12 5 432

Serdeczny, O., Adams, S., Baarsch, F., Coumou, D., Robinson, A., Hare, W., Schaeffer, M., Perrette, M., & Reinhardt, J. (2017). Climate change impacts in Sub-Saharan Africa: From physical changes to their social repercussions. *Regional Environmental Change*, *17*, 1585–1600.

Serrat, O., & Serrat, O. (2017). The sustainable livelihoods approach. *Knowledge solutions: Tools, methods, and approaches to drive organizational performance*, 21–26.

Sharma, A., Shukla, A., Attri, K., Kumar, M., Kumar, P., Suttee, A., Singh, G., Barnwal, R. P., & Singla, N. (2020). Global trends in pesticides: A looming threat and viable alternatives. *Ecotoxicology and Environmental Safety*, *201*, 110812.

Shire, C. (2016). Men don't go to the moon: Language, space and masculinities in Zimbabwe. In *Dislocating masculinity* (pp. 154–164). Routledge.

Siebert, J., Sünnemann, M., Auge, H., Berger, S., Cesarz, S., Ciobanu, M., Guerrero-Ramírez, N. R. & Eisenhauer, N. (2019). The effects of drought and nutrient addition on soil organisms vary across taxonomic groups, but are constant across seasons. *Scientific Reports, 9*(1), 639.

P Smith JI House M Bustamante J Sobocká R Harper G Pan TA Pugh 2016 Global change pressures on soils from land use and management Global Change Biology 22 3 1008 1028

Taylor, R. L. (1992). *Entertaining with insects; or: The original guide to insect.*

Tchibozo, S., Van Huis, A., & Paoletti, M. G. (2005). Notes on edible insects of South Benin: A source of protein. *Ecological implications of Minilivestock: Potential of insects, rodents, frogs, and snails*, 245–250.

LS Teffo 2006 Preliminary data on the nutritional composition of the edible stink-bug, *Encosternum delegorguei* Spinola, consumed in Limpopo province, South Africa South African Journal of Science 103 434 436

Ulber, B., Williams, I. H., Klukowski, Z., Luik, A., & Nilsson, C. (2010). Parasitoids of oilseed rape pests in Europe: Key species for conservation biocontrol. *Biocontrol-based integrated management of oilseed rape pests*, 45–76.

A Huis van 2003 Insects as food in sub-Saharan Africa Insect Science and Its Application 23 163 185

Van Huis, A., & Oonincx, D. G. (2017). The environmental sustainability of insects as food and feed. A review. *Agronomy for Sustainable Development*, *37*, 1–14.

Youdeowei, A., Ezedinma, F. O. C., & Onazi, O. C. (2000). *Introduction to Tropical Agriculture*. Longman. London.

Zacharia, J. T. (2011). Ecological effects of pesticides, pesticides in the modern world-risks and benefits, Dr. Margarita Stoytcheva (Ed). ISBN
WJ Zikmund 2003 Business research methods 7 Oklahoma State University
Zimbabwean Vulnerability Assessment Committee (ZIMVAC) report. (2012). Harare.

CHAPTER 8

Climate Change and Its Effect on the Mental Health of Farmers in Trinidad and Tobago

Sadia Moore, Glenda Hinkson, and Edward Clarke

Introduction

The Caribbean region has been identified as one of the most vulnerable areas to climate change (Lal et al., 2002; Mimura et al., 2007), faces profound challenges due to its intricate reliance on climate patterns. This dependence extends across various facets of Caribbean economies, lifestyles, and operational cycles, rendering them highly susceptible to climate fluctuations (Taylor et al., 2012). From intensified hurricanes and coastal erosion to coral reef degradation, the region contends with a host of environmental threats, jeopardizing its socioeconomic stability

S. Moore (✉) · G. Hinkson · E. Clarke
University of the Southern Caribbean, Maracas, St. Joseph, Trinidad and Tobago
e-mail: moores@usc.edu.tt

G. Hinkson
e-mail: hinksong@usc.edu.tt

E. Clarke
e-mail: eclarke@usc.edu.tt

D. D. Joseph et al. (eds.), *Climate Change*,
https://doi.org/10.1007/978-3-031-83037-2_8

and the well-being of its inhabitants. Amid these adversities, the toll on mental health looms large, a critical yet often overlooked consequence that profoundly affects vulnerable Caribbean communities. Farming in Trinidad epitomizes these challenges, grappling with limited arable land, reliance on oil and gas, and the adverse impacts of climate change, including water shortages, soil erosion, and land fires. With diverse farming scales, including large-scale commercial ventures and small personal plots often leased from the government to sustain food production, Trinidadian farmers navigate a landscape deeply entwined with climatic rhythms and uncertainties. The seasonal dynamics of Trinidad's climate, characterized by distinct dry (December–April) and wet (May–November) periods, alongside temperature fluctuations, dictate crucial agricultural practices such as planting and harvesting (Mimura et al., 2007; Taylor et al, 2002). These practices, deeply ingrained in the fabric of Trinidadian agriculture, underscore the profound influence of climate on the farmers' livelihoods. This study endeavors to shed light on the intricate nexus between climate change and mental health in the Caribbean, focusing mainly on the experiences of farmers whose livelihoods are intricately tied to climatic conditions. By exploring the region's unique environmental vulnerabilities and examining its diverse populations' psychological repercussions, this research aims to foster a deeper understanding of this pressing issue. Moreover, it underscores the urgent need for holistic approaches to address the Caribbean's environmental stressors, recognizing the region's intertwined nature of climate, mental health, and socioeconomic resilience.

The intricate link between environmental degradation and mental health is a critical area of study, particularly concerning its impact on the psychosocial well-being of farmers. The vulnerability of the agriculture sector to a variable and changing climate is widely recognized (EPA Caribbean Environmental Programme, 2022), with observable shifts in traditional weather patterns already leading to production and productivity challenges. This shift toward a new and "unfamiliar" regional climate regime, characterized by warmer temperatures, erratic rainfall patterns, intensified hurricanes, and rising sea levels, poses significant threats to regional economies, with agriculture and rural farming communities at the forefront of impact (Rhiney, 2018). Ten farmers in Trinidad were interviewed to delve deeper into these challenges and elucidate their farming obstacles and the resulting mental health stressors exacerbated by climate change. Through this lens, the psychological toll on farmers was explored, considering personal and employment disparities that influence individuals' capacity to cope with environmental change-induced mental health consequences. Subsequently, the implications of these findings for policy, practice, and research, stressing the necessity of integrating mental health considerations into climate change adaptation and mitigation strategies will be discussed.

Globally and locally, climate change poses a formidable challenge to farming systems. Effective adaptation and mitigation strategies can be facilitated by locally tailored outreach programs framed within farmers' perceptions and values. Moreover, the multifaceted agricultural, economic, social, and environmental challenges necessitate research and engagement efforts capable of identifying and adapting to farmers' dynamic priorities and perceptions. By fostering interdisciplinary collaboration and leveraging community-based approaches, we can enhance resilience, promote psychological well-being, and cultivate sustainable futures for the Caribbean and beyond.

Climate Change in the Caribbean

The impact of human activities on the Earth's climate and the subsequent increase in average global temperatures can be attributed to the release of greenhouse gases such as carbon dioxide, methane, and nitrous oxide into the atmosphere (The National Aeronautics and Space Administration [NASA], 2024). In the last twenty years the oceans heat content has increased significantly with the year 2022 regarded as the fifth or the sixth warmest year on record (World Meteorological Organisation [WMO], 2022). The warming of the oceans has led to the ice sheets melting at an accelerated rate resulting in an increase in sea levels of 4.62 mm per year between 2013 and 2022, reaching a new record high in 2022.

Climate change affected 284 million persons globally between 2008 and 2020 according to the 2021 Global Report on Internal Displacement (GRID) of the Internal Displacement Monitoring Centre (IDMC). Drawing from the facts presented by (IDMC, 2021), populations have become displaced due to natural disasters such as floods, storms, wildfires, droughts, extreme temperatures, and landslides. The United Nations Educational, Scientific and Cultural Organisation reported that in 2019 alone 23.9 million people globally were displaced due to weather-related disasters (UNESCO, 2019). The socioeconomic characteristics and geographic location of Latin America and the Caribbean (LAC) region make it highly susceptible to climate change and displacement impacts. The Caribbean, generally classified as comprising small island developing states (SIDS) is particularly vulnerable to the adverse effects of changing weather patterns and rising sea levels due to its unique combination of social and economic factors, and geographical positioning (United Nations, 2024). The results of climate change have caused the occurrence of severe tropical cyclones, storm surges, prolonged droughts, shifting patterns of precipitation, rising sea levels, and coral bleaching are some of the significant impacts associated with intense tropical cyclones in small island developing states (Intergovernmental Panel on Climate Change [IPCC], 2014).

According to WMO (2022), LAC is witnessing a growing intensity of extreme weather events and climate shocks. These phenomena are increasingly severe and impactful, posing significant challenges to the region. According to current trends, the proportion of regional gross domestic product (GDP) is expected to rise significantly from 5 % in 2025 to over 20 %by 2100 (Thomas et al., 2020). Climate change is not limited to environmental consequences but also has far-reaching effects on human settlements, public infrastructure, the overall health and welfare of populations, access to clean water and food sources, and communities' economic and cultural fabric (Stern, 2007).

In the context of Trinidad and Tobago, it is notable that although the country's contribution is less than one percent of the global greenhouse gas emissions, the fossil fuel-dependent economy contributes to a significant level of greenhouse gas (GHG) emissions per individual (Ritchie et al., 2020). From 1946 to 2019 the island has warmed by about 2.1 °C making it approximately two and a half times the global average (Trinidad & Tobago Meteorological Service, 2019). It is worth mentioning that seven out of the last ten years have been classified as part of the twelve warmest years on record. Additionally, intense precipitation occurring within 24 hours or less have been attributed to flooding whereas substantial and prolonged rainfall can lead to extensive flooding across vast areas of the nation. Notwithstanding, despite small island developing states (SIDS) being particularly at risk of the impacts of climate change, the existing comprehension of how small islands respond to climate change is severely deficient (Duvat et al., 2017). According to (Mycoo & Donovan, 2017; and Robinson & Dornon, 2017), these concerns are paramount given the islands' unique biodiversity of the islands and data deficiency.

Geographic Makeup

Situated at the southernmost point of the Caribbean archipelago, Trinidad and Tobago comprises two islands, forming a unique country in the region; it shares maritime borders with Venezuela, Grenada, and Barbados (Sweeney, et al., 2010). The geological evidence suggests that the two islands were previously connected to the South American continent, forming a contiguous landmass in the past. Vestiges of this relationship endure in the chain of petite islands near the north-western tip of Trinidad and in its plant and animal life, which resemble to those found in Venezuela. The island's geographical features of the island play a significant role in shaping its climate, with influences from its island characteristics, proximity to the equator, mountain chains, oceans and the prevailing northeast trade winds. These factors collectively contribute to the unique weather patterns experienced on the island, referred to as the dry and wet seasons.

Response to Climate Change

UNESCO (2022) considers the Caribbean as the starting point for the emergence of global climate. Additionally, the WMO (2021) highlights implications for ecosystems, and the security of food and water, human health, and poverty, which the significant and wide-ranging consequences of climate change have been underscored. Governments have begun acting in their respective countries through various programs and initiatives. Barbados developed Roofs to Reefs in 2021. In 2020, Jamaica achieved a significant milestone by becoming the inaugural Caribbean nation to present a revised climate action plan that establishes more ambitious goals for decreasing emissions in energy and land-use domains (International Institute for Sustainable Development [IISD], 2020). Additionally, Dominica's government in 2018 established the Climate Resilience Execution Agency to bolster the island's ability to respond to climate events (Robinson & Butchart, 2022).

Trinidad and Tobago stood out as one of the first countries in 1994, to officially endorse and approve the United Nations Framework Convention on Climate Change through signing and ratification. In 2015, Trinidad and Tobago implemented a carbon reduction strategy and subsequently signed the Paris Agreement committing to an overall cumulative emissions reduction a year later (Ministry of Planning & Development,

2021). Regionally, efforts have been made to address the issue by creating the Caribbean Community Climate Change Centre (CCCCC) in 2002 and the Climate Governance Initiative for the Caribbean project (CARICOM, 2021–2024). Despite these efforts, climate finance is another major challenge for Caribbean nations (Roy, 2023).

Climate Change and Mental Health

The impact of climate change is widely recognized as a global concern, with ominous findings indicating the severity of its consequences. However, it is essential to acknowledge that the Caribbean region is particularly vulnerable and is expected to experience a disproportionately more significant impact from these effects (Rise et al., 2022). According to (Hellerstedt et al., 2017) the twenty-first century will witness the persistent and alarming impact of climate change on public health, posing a significant threat that demands urgent attention. The impact of climate change is further compounded by the Caribbean's inability to protect and preserve health care; this is further exacerbated by a lack of thorough investigation into the actual health consequences resulting from climate change and a lack of research on preventative methods and preparedness on health impacts (Organisation for Economic Co-operation and Development [OECD], 2012).

Macpherson and Akpinar-Elci (2013) postulates that the void of research on the issue of climate change in the Caribbean is of grave concern considering the region is on the frontlines of the phenomenon with increased levels in temperatures, floods, drought, rising sea levels, and more frequent and intense storms. The contemporary era is characterized by the prominence of climate change and mental health issues, which are two significant challenges facing society. Research has associated climate changes with physical health problems, whereas the psychological effects of this phenomenon still need to be more adequately comprehended. Climate change has brought about an emerging psychological phenomenon termed "climate anxiety" or "eco-anxiety," which has been described as a "chronic fear of environmental doom" (Clayton et al., 2017, p. 68).

Every-Palmer et al., (2016) study was the first to focus on the association between climate change and its impact on mental health. Findings concluded that the prevalence of post-traumatic stress disorder, depression, anxiety, and substance use disorders is amplified by the occurrence of extreme weather events, such as floods, droughts, heat waves, wildfires, and storms. Hayes et al. (2018), also acknowledged that mental health problems such as grief, survivor guilt, hopelessness, difficulties in social relationships, and suicidal ideation can also become apparent in individuals. Kabir (2018) discovered that individuals whose means of living are impacted by climate change undergo emotional strain due to the importance of employment for their overall health and happiness. The overarching threats of a changing climate can also present barriers to care that can worsen the existing shortcomings within the healthcare system, mainly when there is a lack of support from high-level policies. These barriers can hinder access to necessary healthcare services and contribute to the persistence of chronic deficiencies in the system (Harris-Glenville & Cloos, 2024). Interestingly, Edwards and Wiseman (2011); Ramsay and Manderson, (2011) found that these unfortunate events can also encourage selflessness, empathy, and a positive outlook, and cultivate a feeling of purpose and individual development.

The acknowledgement of the impending perils and current ramifications of climate change gives rise to obstacles in terms of emotional and social well-being (Fritze et al., 2008). Albrecht et al., (2007) recognition of the psychological impact caused by the understanding of the significant issues posed by global warming has been noted. Albrecht et al., (2007 as cited in Hayes et al., 2018) acknowledged: that this awareness contributes to psychoterratic syndromes. Psychoterratic syndromes include phenomena such as ecoanxiety, ecoparalysis, and solastalgia. Ecoanxiety refers to the anxiety people face from constantly being surrounded by the wicked and threatening problems associated with a changing climate. Ecoparalysis refers to the complex feelings of being unable to take effective action to mitigate climate change risks significantly. Solastalgia refers to the distress and isolation caused by the gradual removal of solace from the present state of one's home environment. (p. 7)

The introduction of this novel set of terms equips individuals with the necessary linguistic tools to delve into the wider psychological consequences associated with the increasing risks posed by climate change.

The Impact of Climate Change on the Agricultural Sector

Agriculture remains a crucial sector in the Caribbean region, playing a significant role in the economy and livelihoods of many people. However, extreme weather events threaten the sustainability and productivity of agricultural activities in the region requiring appropriate policies and legislation (Ganpat & Isaac, 2014, p. 13). In the publication released by (FAO, 2017), the agricultural and food sectors have experienced significant losses due to these disasters, with a noticeable upward trend in the amount of damage incurred. Henry (2012) further explains that the Caribbean region's susceptibility to natural calamities such as heat waves and droughts wreaks havoc on crops and agriculture's fundamental aspects, leaving a lasting impact on the industry.

Warmer weather conditions can lead to shifts in where and when weeds, insect pests, and pathogens grow, impacting their overall distribution patterns (Levy & Roelofs, 2016). The alterations have the potential to impact the well-being and security of farm laborers and individuals in various professions who regularly interact with wildlife. Agricultural workers are highly vulnerable to health problems caused by occupational exposure to toxic substances, especially pesticides, due to the misuse of personal protective equipment and a lack of knowledge about its proper use (de-Assis, et al., 2021). Similarly, agricultural workers in the Caribbean region could also face health hazards from climate-drives challenges. The escalation of these diseases would not only result in higher rates of sickness and mortality, but still, it would also impose a more significant strain on the healthcare infrastructure within the area. The climate sensitivity of the region which relies on economic sectors like agriculture, is heavily dependent on seasonal rainfall for water supply. According to Moraes et al., (2023) the Caribbean region may experience a decrease in the limited water supply because of higher evapotranspiration rates. This could further exacerbate the region's challenges regarding water scarcity and availability. Due to water dependency in agriculture, the upcoming drying trends are expected to directly impact on this industry, mainly affecting smallholder farmers who rely heavily on the quantity and timing of yearly rainfall (Bates et al., 2008; Curtis et al., 2014).

Methodology

The qualitative intrinsic case study methodology was chosen for its suitability in examining contemporary phenomena within natural settings (Yin, 2009). Additionally, Creswell (2007) advocates for case studies when there are identifiable cases with clear boundaries, aiming to offer profound insights into the cases or facilitate comparisons among them. This methodology is particularly apt for exploring specific groups or events, as emphasized by Hancock and Algozzine (2016), and Stake (1995) suggests its efficacy in gaining insights into particular cases.

Data analysis was conducted using thematic analysis, allowing for identifying and exploring recurring patterns and themes within the data.

Fifteen farmers from Trinidad consented to participate in the study. While some were unavailable and could not be reached to schedule interviews, one participant commenced the interview but discontinued midway, despite follow-up attempts. Ultimately, nine farmers completed the interviews, reaching a saturation point where findings and responses converged.

Several participants graciously welcomed the researchers to their farms, providing tangible evidence of climate change's impact on the size of their produce. All participants signed consent forms, ensuring ethical compliance throughout the study. Approval was granted from the University of the Southern Caribbean, Office of Research and Innovation, with the IRB#: USC-RE-REC-002–2024-332.

Interviews, lasting approximately 45–60 minutes each, were conducted face-to-face or via Zoom, and accommodating participants' preferences. Personal interviews and direct observation served as the primary methods for data collection, fostering rich, nuanced insights into the farmers' experiences and perspectives. The participants were quite forthcoming and candid about their experiences and mental well-being. This can lead to more genuine and insightful sharing. Their willingness to share suggests they felt their experiences and perspectives were valued and allowed for gaining deep insights. Overall, these characteristics suggest a positive and productive interaction, providing valuable insights into the participants' mental health and well-being.

Findings

The primary objectives of this study were to, (1) understand the challenges farmers face due to climate change and identify potential strategies for adapting agricultural practices to mitigate these challenges, and (2) identify the prevalence of stress, anxiety, depression, and other mental health challenges among farmers and understand how these changing climate conditions may exacerbate these issues. Furthermore, it will uncover potential coping mechanisms and interventions to support the mental well-being of farmers facing climate-related challenges. The research question that guided this exploration was: How does climate change affect the agricultural sector in Trinidad and Tobago, and what are the associated impacts on the farmers' economic stability and mental health of farmers? Unearthing the answer to this question is fundamental to understanding the psychological effects of climate change on the mental health of farmers in Trinidad.

This study used thematic analysis which required the transcription of interviews and followed coding stages. The authors carefully read transcripts to identify themes and sub-themes. This section commences with a table detailing the demographic information of each participant. In Table 1 outlined below, each participant is identified by an alphanumeric code, to ensure their identities remain confidential. Table 8.1 depicts the demographic information of the nine (9) participants, detailing their age range, ethnicity, number of years they have been farming, the size of their farms and their locations.

Themes and Sub-themes

Three themes were identified emerging from the study; knowledge, crop yield and livelihood, and mental health, along with subthemes; Climate Change Impact, Adjusted Planting & Harvesting Schedules, Farmers' Concerns, Feelings, Behavioral Changes, and Coping Strategies, which are discussed next.

Theme One: Acknowledgment of Climate Changes

The participants in this study agreed that Trinidad has been impacted by climate change, and changes in the weather adversely affected their farming. They all attributed the adverse effects to climate change, and

Table 8.1 Demographic information

	Age range	*Ethnicity*	*Years in farming*	*Size of farm*	*Farm location*
P1:	62	Indo-Trinidadian	50 yrs	10, 16, 30 acres	Costal, rural, urban (Manzanilla, Oropouche, Piarco)
P2:	32	Mixed-Trinidadian	12 yrs	2.5 acres	Rural (Maracas valley, St. Joseph)
P3:	63	Afro-Trinidadian	25 + yrs	8 acres	Rural (Buenos Aryes Point Fortin
P4:	58	Mixed-Trinidadian	40 yrs	25 acres	Rural (Princes town, south central Trinidad
P5:	33	Mixed-Trinidadian	13 yrs	4 acres	Rural (Arima)
P6:	58	Indo-Trinidadian	35 yrs	3 acres	Rural (Carlson field, Chaguanas)
P7:	63	Indo-Trinidadian	34 yrs	5 acres	Rural (St Mary's village Moruga, Princes town)
P8:	72	Indo-Trinidadian	40 yrs	5 acres	Rural (sixth company village, Princes town)
P9:	65	Afro-Trinidadian	45 yrs	2.5 acres	Rural (La Baja, St. Joseph more inland)

more than half of the participants noted changes in the weather over the past ten years.

One participant shared:

> Years ago in the dry season you did not have to wet the plants like you have to now because its hotter and the dry season is longer. Now the hotter and longer dry season has caused a lot more bush fires. In the Maracas St. Joseph area there are a lot of bushfires, and the hot season increases the frequency of bush fires. I had a 40-year-old mango tree that was destroyed by bush fires.

Another participant shared:

> There's more sun, and many dry spells during the rainy season. When there's rain it's in excess which affects the plants. The sun and rain are not as seasonal as they once were so you can expect rain and sun anytime.

Sub-theme: Climate Change Impact

The change in the weather patterns has further affected participants' farming practices. One participant shared:

> The way crops grew it's not growing like that anymore, the extreme heat and extreme rain, so you have to use chemicals because of the soil. However, south soil is more fertile because it was never cultivated sugarcane. Lots of chemicals are used on lands once used for sugar cane farming.

Another participant shared:

> In 2028 there was a flood in Trinidad that changed the pattern of the river where I farm and that affected my garden. Over the last three year I started noticing more of the African snails and they feed on the crops at night.

Yet another participant shared:

> …the rainy season has been pushed back, for instance, you hardly get April showers now and sometimes it takes as long as June.

Theme Two: Crop Yield and Livelihood

All of the participants in this study expressed that the changes in the climates have adversely affected their crops' yield, affecting their livelihood. They expressed that the health of their crops have been affected, their harvest has dropped and in some instances there has been a one hundred percent loss, hence no crops to harvest. As a result of reduced crop yield and poor plant health, all the farmers said that their livelihood has been affected in several ways.

One participant shared:

> This year (2024) my plantains production dropped by approximately 50% and the eggplants dropped by 70% because it was hard to get water to the land. Even my cassava had a shorter shelf life…

Another participant shared:

> …the coconuts are now more affected as the sun is getting hotter and the coconut trees needs a lot of water, so now I get a smaller coconut when it's time to harvest, which means less production and this applies to everything else that I planted.

Another participant shared:

> …very badly because when the Lisbon yam has declined vastly because of the weather and disease. The sun has affected the yield of the crop in terms of the size of the yams. The yams are now smaller from 10 lbs per hill to 3lbs a drastic change.

Yet another shared:

> …due to lack of water which helps the chemicals to work when harvesting I get sometimes about half of what I invested so I had to find other ways of taking care of my family, so I started doing barbering to help take care of my family.

Another participant shared:

> In the early 80s and 90s I made more money due to greater yield and less effort now in the 2000s I have to put much more effort and the yield is less. The amount of money that I get now has decreased, so I got a tax

badge to provide taxi services part time to compensate my income and help my family.

Sub-theme: Adjusted Planting and Harvesting Schedules

All of the farmers agreed that they have had to adjust their planting and harvesting schedules in response to the changing weather conditions and they also had advice to offer to other farmers who are also dealing with climate change-related challenges.

One participant shared:

> ...sometimes I have to pick crops earlier if not its expensive to buy pesticides and fungicides to protect the plant. Sometimes I have to plant double to ensure that we get the amount needed weekly due to customer demands. This also requires more space.

Other participant views include the following:

View one:

> I have to wait longer to plant coconut trees I do it when the rain comes. In 2023 part of my coconut farm caught fire due to natural bush fire so I had to make a terrace to prevent the bush fire from crossing over to my farm.

View two:

> We are currently at the mercy of the weather so when the rain comes you have to take full advantage of the opportunity to plant my crops.

View three:

> I don't plant all my seedlings at the same time, I spread it over a few months, for instance if I have to plant 300 seedlings I do it in 100s, to ensure that I get at least some returns.

View four:

> Other farmers need to education themselves on the impact of climate change and learn how to best combat the effect because knowledge is

important for change and betterment. Farmers who are not members of markets should get membership and those who don't have farmers badge should be legalized to get access, therefore there's a need for land regularization for farmers to get farmer badges.

View five:

> Focus on finding places to farm where there is irrigation. Farmers can diversify and plant other crops. I have now started to plant fruit trees, coconuts, lime and avocado.

View six:

> …we have implemented a drip waist system that helps to control the amount of water and food plants get which saves water and nutrients. We've also invested in green houses to protect our crops from the sun and gotten grow boxes to help control run off water.

Sub-theme: Farmers' Concerns

When asked if they have any concerns about the long-term sustainability of their farming operation in the face of continued weather changes, 8 of the 9 participants said that they did have concerns and shared their concerns such as management, fluctuating river levels, growing intensity of the heat, and the rising costs of pesticides, herbicides, fungicides, and fertilizers.

One participant shared:

> …with the current weather pattern being more hot I am concerned that I will lose a lot of plants because the plants will die out. The cost will go up because there will be a decrease in production. Also pests and diseases have multiplied.

Another participant shared:

> …because of the uncertainty of the weather I felt concerned that I would not be able to provide for my family, so I diversified and started poultry farming to balance things off economically, because there were times when the weather did not allow any harvesting.

Theme Three: Mental Health

All the participants believe that their mental health has been affected. They all expressed feeling discouraged, frightened, sad, fearful, worried, frustrated, hopeless, depressed, and stressed because of the change in weather patterns. Many expressed that some of these feelings occur daily, weekly, or monthly. The participants acknowledged experiencing distress and emotional challenges, yet no visual representations of these emotions were observed during the interviews. Despite the impact on their mental health, there were no emotional outbursts or other visible signs of their emotional state.

One participant shared:

> I feel frightened and sad that I can't do what I love.

Other participants' views include the following:

View one:

> ...daily other farmers call me and tell me their concerns, this makes me sad what the others farmers like myself have to go through."

View two:

> ...this is so frustrating because I'm making less money and have to take care of my household and this is very worrisome.

View three:

> When I plant short crops I feel worried and fearful daily but when it's a long term crop I feel this way monthly.

View four:

> I feel worried because if it's like that now how would it be in two years. It frustrating because you don't have much control or save. I have two choices to stay in the garden or forget it.

View five:

> I see plants as people so I feel confined that I cannot attend to my plants the way I would want to attend to them, this is depressing.

View six:

> I feel stressed because when you make an investment in a nursery its expensive, and you're depending on the rain and get no rain you worry about your investment.

Sub-theme: Behavioral Changes

When asked if there have been any changes in their behavior due to how they felt, fifty percent of the participants said yes and that this affected their relationships with family and friends.

One participant shared:

> ...when there are bad days in the garden I tend to act out with my family.

Here are the views of other participants:

View one:

> I don't sleep like I used to because I worry about my crops.

View two:

> I tend to get angry more often and have sleepless nights thinking about my crops.

View three:

> ...I don't have extra time to spend with my friends like I used to and they don't always understand that. With my family when I have a bad day on the farm I get annoyed when my family approach me but this happens once in a while.

View four:

> …verbal communication is sometimes poor with my family and I get more quiet and withdrawn.

View five:

> …sometimes I speak to my wife roughly, get more short-tempered and get more quiet…

Sub-theme: Coping Strategies

The participants shared various strategies that they practiced helping them cope and their thoughts on what the government can do to help alleviate the mental health of farmers in Trinidad.

The views of some of the participants are:

View one:

> I exercise, watch movies, read and surf YouTube for suggestions on planting.

View two:

> I have gone for counselling, spend time alone in my garden and I listen to lot of gospel and country & western music. Also my religion has taught me to depend on God and have faith.

View three:

> I practice yoga weekly which has really helped me to learn to be at peace and accept the realities of life and wrap my mind about it. This has given me much peace.

View four:

> Despite my moods, my family is still very supportive. We have a lot of family traditions and do things together on weekends, which helps a lot.

View five:

> The government can create a system where they can talk and vent to handle emotional problems. Give them access to mental health professionals.

View six:

> ...The government can provide pension plans for farmers and give them crop insurance.

View seven:

> ...The government can provide programs to help farmers deal with their mental health and also provide counselling services for farmers.

View eight

> ...The government can find ways to regulate the market when the prices go down in the market.

View nine

> *...The government can provide extension officers who are trained in counselling or can refer farmers to therapist provided by the government."*

View ten

> A farmer has to be registered to get government subsidies but many people do not have proper land tenure so though they are farming they cannot access the subsidies. If farmers are producing they should be given subsidies, therefore the subsidies formula needs to be revisited.

Discussion

This study explored and discussed the mental health impacts of climate change by utilizing informal semi-structured interviews and field observations and informal interviews with farmers in Trinidad. In Trinidad, many small and medium-sized farms have been adversely impacted by climate change. Farmers are challenged to decide whether to remain in farming or seek other sources of income to support their families.

Regarding crop adaptation and adjustment, eight of the nine participants admitted to adjusting their planting and harvesting schedules in response to changing weather conditions. Similarly, results from a study in Ghana, indicated that farmers used 17 adaptation strategies to adapt to climate change. The most common strategies identified were change of planting days, crop diversification, use of resistant varieties, and monitoring weather forecasts on the radio. Furthermore, the factors that influenced the choice of adaptation strategies were experience in farming, household size, and perceptions about the impact and intensity of climate change (Aidoo, 2021).

Instead of research studies, newspaper articles, like one by Christopher (2024), highlight the urgency for farmers to adapt amid climate change concerns. The Marivista Institute for Agriculture, Training and Development (MIATD) convened a seminar titled "Climate Change and Paramin Agriculture" at the Maraval Community Centre, where they addressed the impact of climate change on farming cycles in the Caribbean, noting that, "Climate change has already started affecting farming cycles in the Caribbean. There is a water shortage in Guyana and Suriname." In the same article, Professor Michelle Mycoo (Trinidad Express, 2023) from the University of the West Indies, specializing in Urban and Regional Planning, noted that anticipated climate effects have already begun, causing extreme heat and unpredictable weather cycles. Mycoo emphasized the challenge for farmers in planning and strategizing due to shifting rain patterns, with some Caribbean nations facing severe drought. "The Caribbean is facing more of that drought effect than the Pacific islands, due to our geographic location. Cuba and Jamaica face even more severe drought than [Trinidad]," noted Mycoo.

For the participant who shared an inability to change the type of crop production, as some farmers have little choice but to persist with (and even invest further in) their current farming endeavor, Patel (2007) proposed a potential solution. For farmers with limited options, Patel suggested they should continue with their current crop and attempt to trade their way out of their predicament. This persistence, while necessary, might be mistaken for resilience.

Climate change-related factors that exacerbate mental health problems will inevitably complicate the effects of cognitive decline and physical health problems that frequently accompany aging for some farmers, who reported mental health changes that can last from monthly to daily to only during crop time. Steptoe et al., (2008) noted that prolonged hopelessness is prospectively linked to depression, substance misuse, other mental health issues, physical illness, and even suicide. Such findings reveal that recently observed patterns of climate change have exacerbated farmers' worries about the weather. The research findings highlight the tightly coupled ecosystem health-human health relationships for family farmers in regions affected by climate change (Ellis et al., 2017). Drought-prone areas are chronically vulnerable to low socioeconomic status and educational attainment. Drought-related deterioration in socioeconomic circumstances in rural areas is linked to higher levels of distress and helplessness (Coelho et al., 2004). Hogan et al., (2008) reported that while farmers reported significantly better well-being than did non-farmers on many aspects of life satisfaction, they also reported less hope for their future.

Various emotions arise from fluctuations of the weather and crop yields, impacting individuals' financial situations, although not all respond with altered behavior. The consequences and hazards of climate change extend across environmental, psychological, agricultural, and economic domains, posing significant threats to health and overall well-being. The participants' openness in describing their experiences in a matter-of-fact manner might indicate a level of acceptance or normalization of their mental well-being or a sign of resilience or a coping mechanism. Ultimately, climate change has been found to compromise physical health and adversely affect mental well-being (Ajator et al., (2021).

Recommendations

Smallholder farmers, predominantly those residing in the Small island developing states (SIDS) within tropical and subtropical regions, like the Caribbean, face precarious conditions exacerbated by climate change. The health repercussions of climate variability perpetuate the poverty cycle among these farmers and pose a significant threat to national and regional food security. Addressing the mental health challenges stemming from climate change is pivotal for enhancing their resilience. In light of this urgency, the following recommendations, synthesized from interviews and participant responses, are proposed to mitigate the health impacts of climate change at both regional and national levels:

All farmers emphasized the critical need for mental health awareness and support. To address this pressing issue, a proactive approach is recommended, including the establishment of a dedicated working committee tasked with organizing symposiums. These gatherings would serve as platforms for farmers to share insights and recommendations while fostering supportive networks. Additionally, implementing accessible systems for farmers to express their emotions and seek assistance for emotional challenges is paramount. This could involve the provision of counselling services and facilitating access to mental health professionals. Currently, there appears to be a gap in governmental support for farmers' mental well-being, highlighting the urgent need for intervention. By deploying counsellors to aid farmers in coping with losses and hardships, significant strides can be made in bolstering their mental resilience. Moreover, integrating counselling training for extension officers or facilitating referrals to government-provided therapists can enhance the accessibility of mental health support. Furthermore, instituting designated days without market activities would afford farmers much-needed time for relaxation and rejuvenation, thereby contributing to the alleviation of mental health strains exacerbated by constant market demands, particularly in instances where farmers lack control over wholesale pricing dynamics.

In times of crop loss, it is imperative that farmers receive prompt access to essential resources and financial aid to mitigate the impact on their livelihoods. This support should encompass provisions such as food aid, financial assistance, and supplementary supplies like seeds, plantings, and fertilizer as necessary to facilitate recovery and future cultivation efforts. To fortify resilience against extreme weather events like flooding and drought, robust regional-scale management and protective infrastructures are indispensable. These measures could include the establishment of insurance mechanisms tailored to farmers' needs, ensuring coverage for crop losses. Moreover, implementing pension plans specifically designed for farmers would provide crucial long-term financial security, further bolstering their resilience in the face of agricultural uncertainties.

To facilitate the adaptation of smallholder farmers to climate change impacts and to safeguard both agriculture and public health, comprehensive guidance and financial support are essential. This assistance should extend to the acquisition of new equipment and the adoption of innovative management practices tailored to climate resilience. Moreover, ensuring market stability is imperative to sustain small farmers' livelihoods and uphold food security. Addressing the mental health challenges

Approval for this research was given from University of the Southern Caribbean Ethics Committee on June 10, 2024. Number USC-RE-REC 002–2024-332.

References

Aidoo DC, Boateng SD, Freeman CK, & Anaglo JN. (2021). The effect of smallholder maize farmers' perceptions of climate change on their adaptation strategies: the case of two agro-ecological zones in *Ghana. Heliyon.* 2021 *7*(11): e08307. https://doi.org/10.1016/j.heliyon.2021.e08307. PMID: 34816030; PMCID: PMC8593439.

Ajator, C., Anabaraonye, B., Ewa, B., Otti, V., & Nwobu, E. (2021). *The Impacts of Climate Change on Mental Health: Adaptation Strategies for Sustainability in Africa., 5*, 67–74.

Albrecht, G., Sartore, G. M., Connor, L., Higginbotham, N., Freeman, S., Kelly, B., Stain, H., Tonna, A., & Pollard, G. (2007). Solastalgia: The distress caused by environmental change. *Australasian Psychiatry: Bulletin of Royal Australian and New Zealand College of Psychiatrists, 15*(Suppl1), S95–S98. https://doi.org/10.1080/10398560701701288

Bates, B. C., Kundzewicz, Z. W., Wu, S., Arnell, N., Burgett, V., & Räisänen, J. (Eds.) (2008). *Climate change and water: Technical paper of the intergovernmental panel on climate change.* IPCC Secretariat. http://www.ipcc.ch/pdf/technical-papers/climate-change-water-en.pdf

Blashki, G. A, Burke, S., Fritze, J. G., & Wiseman, J. (2008). Hope, despair and transformation: Climate change and the promotion of mental health and well-being. *International Journal of Mental Health Systems, 2,* Article 13. https://doi.org/10.1186/1752-4458-2-13

CARICOM. (2021–2024). *Caribbean Community Climate Change Center (CCCCC).* https://caricom.org/institutions/caribbean-community-climate-change-centre-ccccc/

Christopher, P. (2024). Farmers urged to adapt, amid climate change concerns. https://www.guardian.co.tt. March 14. Retrieved April 29, 2024

Clayton, S., Manning, C., Krygsman, K., & Speiser, M. (2017). *Mental health and our changing climate: Impacts, implications, and guidance.* American Psychological Association and ecoAmerica. https://tribalclimateguide.uoregon.edu/literature/mental-health-and-our-changing-climate-impacts-implications-and-guidance

Coelho, A., Adair, J., & Mocellin, J. (2004). Psychological responses to drought in northeast Brazil. *Interam J Psychol., 2004*(38), 95–103.

Creswell, J. W. (2007). *Qualitative inquiry and research design: Choosing among five approaches* (2nd ed.). Sage.

Curtis, C. J., Battarbee, R. W., & Monteith, D. T. (2014). The future of upland water ecosystems of the UK in the 21st century: A synthesis. *Ecological Indicators, 37*(Part B), 412–430.

Curtis, G. P., Slay, M. C., Harris, L. N., Tyukavina, A., & Hansen, C., M. (2018). Classifying drivers of global forest loss. *Science*, 61(6407), 1108–1111. https://doi.org/10.1126/science.aau3445

de-Assis, M.P., Barcella, R. C. Padilha, J. C., Pohl, H. H., & Krug, S. B. F. (2021). Health problems in agricultural workers occupationally exposed to pesticides. *Revista brasileira de medicina do trabalho: publicacao oficial da Associacao Nacional de Medicina do Trabalho-ANAMT, 18*(3), 352–363. https://doi.org/10.47626/1679-4435-2020-532

Duvat, E., Magnan A. K., Wise, R. M., Hay, J. E., Fazey, I., Hinkel, J., Stojanovic, T., Yamano, H., & Ballu, V. (2017). Trajectories of exposure and vulnerability of small islands to climate change: Small islands and climate change. *Wiley Interdisciplinary Reviews: Climate Change*, *8*(6). https://doi.org/10.1002/wcc.478

Ebi, K. L., Lewis, N. D., & Corvalan, C. (2006). Climate variability and change and their potential health effects in small island states: information for adaptation planning in the health sector. *Environmental Health Perspectives*, *114*(12), 1957–1963. https://doi.org/10.1289/ehp.8429

Edwards, T., & Wiseman, J. (2011). Climate change, resilience and transformation: Challenges and opportunities for local communities. In: Weissbecker (Ed.), *Climate change and human well-being* (pp. 185–200). Springer.

Ellis NR, & Albrecht GA. (2017). Climate change threats to family farmers' sense of place and mental wellbeing: A case study from the Western Australian Wheatbelt. *Soc Sci Med.* 175, 161–168. https://doi.org/10.1016/j.socscimed.2017.01.009. Epub 2017 6. PMID: 28092757.

EPA Caribbean Environmental Programme. (2022). Climate Change Impacts in the Caribbean: *A regional perspective.*

Every-Palmer, S., McBride, S., Berry, H., & Menkes, D. B. (2016). Climate change and psychiatry. *The Australian and New Zealand journal of psychiatry*, 50(1), 16–18. https://doi.org/10.1177/0004867415615946

FAO. (2017). *Migration, agriculture and climate change—Reducing vulnerabilities and enhancing resilience.* Rome (2017). I829EN/1/12.17.

Fritze, J. G., Blashki, G. A., Bruke, S., & Wiseman, J. (2008). Hope, despair and transformation: Climate change and the promotion of mental health and wellbeing. *International Journal of Mental Health System, 2*(1), 13.

Ganpat, W., & Isaac, W. (2014). Impacts of climate change on food security in small Island developing States. *IGI Global*. https://doi.org/10.4018/978-1-4666-6501-9

Hancock, D. R., & Algozzine, B. (2016). *Doing case study research: A practical guide for beginning researchers*. Teachers College Press.

Harris-Glenville, F., & Cloos, P. (2024). I think they should give primary health care a little more priority. The primary health care in Caribbean SIDS: what can be said about adaptation to the changing climate? The case of Dominica-a qualitative study. *BMC primary care*, *25*(1), 65. https://doi.org/10.1186/s12875-024-02311-w

Hayes, K., Blashki, G., Wiseman, J., Burke, S., & Reifels, L. (2018). Climate change and mental health: Risks, impacts and priority actions. *International Journal of Mental Health Systems, 12*, 28. https://doi.org/10.1186/s13033-018-0210-6

Hellerstedt, W., Toscano, W., Raab, K., Tupper, N., Kumar, V., Josiah, J. S., Grossman, E., Dorr, F., Moran, C., Sehgal, H. & Karpatne, A. (2017). *Climate change and public health.* Center for Leadership Education. https://www.researchgate.net/publication/317569526_Climate_Change_and_Public_Health

Henry, F., J. (2012). Food and nutrition research in the Caribbean. *The West Indian Medical Journal*, *61*(4), 338–350. https://doi.org/10.7727/wimj.2012.149

Hogan A, Polidano C, Russell J, & Stakelum P. (2008). The social wellbeing of rural Australians: An analysis of the household, income and labour dynamics in Australia (HILDA) Longitudinal Dataset. Canberra, Australia: Bureau of rural sciences.

Internal displacement monitoring centre (IDMC). (2021). *Global report on internal displacement 2021.* https://www.internal-displacement.org/global-report/grid2021/

International Institute for sustainable development. (2020). *Jamaica's Updated NDC Incorporates Land-use Change and Forestry Sector*. https://sdg.iisd.org/news/jamaicas-updated-ndc-incorporates-land-use-change-and-forestry-sector/

Intergovernmental panel on climate change. (2014). *Climate change 2014: Impacts, adaptation, and vulnerability. Regional aspects. Contribution of working group II to the fifth assessment report of the intergovernmental panel on climate change*. Cambridge University Press

Kabir, S. M. S. (2018). Psychological health challenges of the hill-tracts region for climate change in Bangladesh. *Asian Journal of Psychiatry, 34*, 74–77. https://doi.org/10.1016/j.ajp.2018.04.001

Lal, M., Harasawa, H., & Takahashi, K. (2002). Future climate change and its impacts over small Island states. *Climate Research, 19*, 179–192.

Levy, B., & Roelofs, C. (2016). Impacts of climate change on workers' health and safety. In: D. V. McQueen (Ed.), *Oxford research encyclopedias*. Oxford University Press.

Macpherson, C., & Akpinar-Elci, M. (2013). Impacts of climate change on Caribbean life. *American Journal of Public Health, 103*(1), e6. https://doi.org/10.2105/AJPH.2012.301095

McSweeney, C., New, M., Lizcano, G., & Lu, X. (2010). The UNDP Climate Change Country Profiles: Improving the accessibility of observed and projected climate information for studies of climate change in developing countries. *Bulletin of the American Meteorological Society, 91*, 157–166.

Mimura, N., Nurse, L., & McLean, R. F. (2007). Small Islands. climate change 2007: Impacts, adaptation and vulnerability. Pp. 687–716 in contribution of working group II to the fourth assessment report of the intergovernmental panel on climate change, edited by M.L. Parry, O.F. Canziani, J.P. Palutikof, P.J. van der Linden, and C.E. Hanson. Cambridge, UK: Cambridge University Press.

Ministry of planning and development. (2021). *First Biennial Update Report of the Republic of Trinidad and Tobago under the United Nations Framework Convention on Climate Change.* https://unfccc.int/sites/default/files/resource/FIRST_%20BUR_TRINIDAD_AND_TOBAGO.pdf

Moraes, S. F., Ramseyer, C., & Gamble, D. (2023). The effects of projected climate change on crop water availability in the U.S. Caribbean. *Journal of Water and Climate Change*, 14 (4), 1176–1191. https://doi.org/10.2166/wcc.2023.398

Mycoo, M., & Donovan, M. (2017). *A Blue Urban Agenda: Adapting to climate change in the coastal cities of Caribbean and Pacific small Island developing states*. https://publications.iadb.org/handle/11319/8264

Organisation for economic co-operation and development. (2012*). OECD environmental outlook to 2050: The consequences of inaction Key facts and Figures*. https://www.oecd.org/env/indicators-modelling-outlooks/oecdenvironmentaloutlookto2050theconsequencesofinaction-keyfactsandfigures.htm

Patel, R. (2007). *Stuffed and starved: Markets, power and the hidden battle for the world's food system*. Portobello Books.

Ramsay, T., Manderson, L. (2011). Resilience, Spirituality and Posttraumatic Growth: Reshaping the Effects of Climate Change. In: Weissbecker, I. (eds) *Climate Change and Human Well-Being. International and Cultural Psychology*. Springer, New York, NY. https://doi.org/10.1007/978-1-4419-9742-5_9

Rhiney, K. (2018). Global Change, Vulnerability, and the Coproduction of Resilience Among Caribbean Farmers. *Small Axe., 57*, 68–80. https://doi.org/10.1215/07990537-7249150

Rise, N., Oura, C., & Drewry, J. (2022). Climate change and health in the Caribbean: A review highlighting research gaps and priorities. *The Journal of Climate Change and Health, 8*. https://doi.org/10.1016/j.joclim.2022.100126

Ritchie, H., Roser, M., & Rosado, P. (2020). *CO_2 and greenhouse gas emissions.* https://ourworldindata.org/co2-and-greenhouse-gas-emissions

Robinson, S., & Dornon, M. (2017). International financing for climate change adaptation in small Island DEVELOPING STATES. *Regional Environmental Change, 17*(4), 1103–1115. https://doi.org/10.1007/s10113-016-1085-1

Robinson, S., & Butchart, C. (2022). Planning for climate change in small island developing states: Can Dominica's climate resilience and recovery plan be a model for transformation in the Caribbean. *Sustainability, 14.* https://doi.org/5089.10.3390/su14095089

Roy, D. (2023). *How the Caribbean is building climate resilience.* https://www.cfr.org/backgrounder/how-caribbean-building-climate-resilience

Schneider, E. (2022, July 3). The Caribbean is 'ground zero' for the global climate emergency. *UN News.* https://news.un.org/en/story/2022/07/1121902

Stake, R. E. (1995). *The art of case study research.* Sage Publications.

Steptoe, A., O'Donnell, K., Marmot, M., & Wardle, J. (2008). Positive affect and psychosocial processes related to health. *British Journal of Psychology, 99,* 211–227.

Stern, N. (2007). *The economics of climate change: The Stern Review.* Cambridge University Press

Sustainable Development and Human Settlements Division of the Economic Commission for Latin America and the Caribbean. (2014). *The economics of climate change in Latin America and the Caribbean Paradoxes and challenges.* https://repositorio.cepal.org/server/api/core/bitstreams/3e136513-76a4-4dd4-b1d5-a118c4e2a45c/content

Taylor, M. A., Enfield, D. B., & Chen, A. A. (2002). The influence of the tropical Atlantic versus. The tropical Pacific on Caribbean rainfall. *Journal of Geophysical Research, 107*(C9), 3127. https://doi.org/10.1029/2001JC001097

Taylor, M. A., Stephenson, T. S., Chen, A. A., & Stephenson, K. A. (2012). Climate change and the Caribbean: Review and response. *Caribbean Studies, 40*(2), 169–200.

The National Aeronautics and Space Administration. (2024). *The Causes of Climate Change: Human activities are driving the global warming trend observed since the mid-20th century.* https://science.nasa.gov/climate-change/causes/

The United Nations Educational, Scientific and Cultural Organization. (2019). *Evaluation of UNESCO's strategy for action on climate change* (2018–2021) "Changing Minds, Not the Climate". https://unesdoc.unesco.org/ark:/48223/pf0000378486?posInSet=3&queryId=N-754be8f6-c6d1-418f-b740-f4555f2c5160

Thomas, A., Baptiste, A., Martyr-Koller, R., Pringle, P., & Rhiney, K. (2020). Climate change and small Island developing states. *Annual Review of Environment and Resources*, *45*(1), 1–27. https://doi.org/10.1146/annurev-environ-012320-083355

Trinidad & Tobago meteorological service. (2019). No Title. https://www.metoffice.gov.tt/

Trinidad Express. (2023, June 1). Hotter weather and El Nino. Sounding the alarm for farmers and food security. *Trinidad Express*. https://trinidadexpress.com/features/local/hotter-weather-and-el-nino/article_f6277488-11ee-9afe-d35aa6980deb.html

UNESCO. (2022, July 3). The Caribbean is 'ground zero' for the global climate emergency. *UN News*. https://news.un.org/en/story/2022/07/1121902

UNESCO, The Right to Education. (2019). *How climate change and displacement affect the right to education*. https://www.unesco.org/en/right-education/climate-change-displacement

United Nations. (2024, 27–30 May). *About small Island developing states* [Conference session] 4th International Conference on SIDS, Antigua and Barbuda. https://sdgs.un.org/smallislands/about-small-island-developing-states

World Health Organization (WHO). (2018). *Mental health and climate change*. WHO document production services

World Meteorological Organization. (2022). *State of the Global Climate 2022*. https://library.wmo.int/records/item/66214-state-of-the-global-climate-2022

Yin, R. K. (2009). *Case Study Research: Design and Methods* (4th ed.). Sage Publications.

PART III

Migration

CHAPTER 9

The Impact of Climate Change Among Displaced Migrants in Small Island Developing States: The Case of Trinidad and Tobago

Camille Huggins

Introduction

Climate change refers to long term shifts in Earth's climate patterns, manifesting as temperature fluctuations, extreme weather events, and human-induced environmental events. These changes adversely affect ecosystems as it diminishes plant life and endangers animal species (Nurse et al., 2014). Small island states play a significant role in creating and pushing the global agenda on climate change because of their small geographical size and constrained resources which make them highly susceptible to significant challenges in mitigating damage (Petzold & Magnan, 2019). The Caribbean is located in the Caribbean Sea and consists of over 700 islands, islets, and cays between North and South

C. Huggins (✉)
The University of the West Indies, St. Augustine, Trinidad and Tobago
e-mail: camille.huggins@uwi.edu

D. D. Joseph et al. (eds.), *Climate Change*,
https://doi.org/10.1007/978-3-031-83037-2_9

America near the equator. Although most of the islands are uninhabited, there are 13 sovereign states and 12 dependent states that are governed by larger developed countries (Baldacchino, 2018). Climate change in the Caribbean has proliferated the concentration of methane and nitrous oxide which emit greenhouse gases from agriculture, livestock, and landfills (Nurse et al., 2014). It also contributes to the deforestation of green spaces (e.g., forest, vegetation) which reduces the planet's capacity to absorb carbon dioxide (Nurse et al, 2014). There is also a rise in average temperatures (e.g., sea levels) which increases heatwaves, droughts, intense cyclones, and hurricanes with ocean warming leading to coastal erosion and inundation (Nurse et al, 2014; Petzold & Magnan, 2019). Adjusting social welfare systems to address the challenges posed by climate change may prove challenging for the 13 sovereign Caribbean states. This chapter is a critical review of one such sovereign state, Trinidad and Tobago's social welfare systems preparedness for climate change considering competing social pressures such as migration.

Caribbean states depend on their natural resources (e.g., land, water, temperature) to sustain their economies (e.g., agriculture, tourism); therefore, they are acutely susceptible to threats of their natural resources. Due to its geographical location, these states face unique challenges as they have relatively small economies, and developmental pressures as they strive to modernize (Petzold & Magnan, 2019). These circumstances complicate emergency assistance and disaster response (Barnett & Waters, 2016).

In Trinidad and Tobago, the social welfare system, which includes institutions such as churches, national government services (e.g., healthcare, criminal justice, and child welfare), United Nations agencies, and non-governmental organizations (NGOs), is ideally positioned to mobilize community action against climate change. However, Trinidad and Tobago has a colonial legacy which involves an inherited top-down governance approach that can be problematic because traditional systems are still prominent in shaping the way natural resources are used (Longepee, 2016). Therefore these entities often do not fully comprehend the severity and consequences of climate issues. As a result, they inadvertently become barriers to the necessary changes needed to address these challenges (Barnett & Waters, 2016). There are also insufficient institutional and financial capacities to implement structural adaptation measures to ensure long-term planning, monitoring, and maintenance (Petzold &

Magnan, 2019). Another issue is the introduction of innovative strategies and mechanisms to address climate change is usually driven by international organizations because of their availability of funding for these initiatives. Approximately 45% of social welfare efforts are funded by the government, while the remaining portion is often supported by bilateral donors, regional organizations, and intergovernmental bodies such as the Caribbean Development Bank and the United Nations (Petzold & Magnan, 2019). Incorporating initiatives funded and driven by external agencies (e.g., international organizations) into a country's internal decision-making processes can be difficult (Petzold & Magnan, 2019).

Given this context, the ongoing challenges of governance, innovation, and resource management in Trinidad and Tobago also highlights the profound impact of historical migration patterns, which have not only shaped the demographic landscape but also continue to influence contemporary societal and environmental vulnerabilities.

Migration

The majority of the population in the Caribbean are descendants of migrants or recent migrants, only the indigenous peoples (e.g., Taino, Caribs, Arawaks) are of the region. Migration served as the catalyst for the development of modern Caribbean civilization, marked by the forced arrival of millions of enslaved Africans who were brought to the region to toil on plantations from 1619 to 1865 (Beckles, 2016). Indentured workers from India, China, and Portugal were bought to the Caribbean when slavery was abolished to also cultivate the land from 1838 to 1920 (Huggins, 2021). When the Spanish settled in the Caribbean, there were only 6% women of Spanish blood living there, so Spanish males procreated with Indigenous women giving birth to a Latin ethnic group which is composed of Indian, Spanish white, Portuguese, and African (Williamson, 1991; Huggins, 2021).

Current migratory patterns are spurred by the Caribbean's economic vulnerability, political instability, and exposure to natural disasters. Many individuals living in the Caribbean opt to migrate to more developed countries such as the United Kingdom, the United States, and Canada. Historically, this trend of migration was spurred by various factors, including the aftermath of World War II and the need for labor in post-war reconstruction efforts, exemplified by the Windrush migration (1948–1971) in the United Kingdom (Basta, 2023). During this period, the British government actively encouraged immigration from its Caribbean colonies, including Jamaica, Trinidad and Tobago, and Barbados (Basta, 2023). In 2020, Trinidad and Tobago is still an emigration country as 330,519 emigrants were recorded in comparison to 78,849 immigrants (Thomas-Hope, 2022).

People may migrate to the Caribbean due to a well-founded fear of persecution, violence, and serious harm based on race, religion, political beliefs, gender, and sexual identification. Persons may be forcibly displaced from their home country and may face ongoing challenges due to resettlement. These people are called *refugees* which have legal rights under international law, primarily through the 1951 United Nations Convention Relating to the Status of Refugees and its 1967 Protocol. These legal instruments define the rights and responsibilities of both refugees and the countries that host them. Refugee status is typically determined through a legal process, and enjoy certain rights and protections under international law, including the right to seek asylum and protection from deportation to a place where their life or freedom may be at risk.

Environmental degradation and natural disasters also displace people, and those under this category are often called "climate refugees." It is estimated that environmental disasters internally displaced 228 million people between 2008 and 2016 (Lister, 2014). Caribbean states are vulnerable to these adverse impacts of climate change with rising sea levels having completely submerged islands affecting potable water supplies and forcing entire communities to relocate to higher ground. The International Organization for Migration estimates that three million residents of low-lying coastal areas are at risk of displacement (Keshen & Lazickas, 2022). The number of environmental migrants crossing international borders is likely to continue to grow as environmental conditions deteriorate (Keshen & Lazickas, 2022).

Another recent migration trend is from South America and Caribbean islands to other Caribbean Islands in hopes of a better economic opportunity. Migration between Caribbean islands is common, with individuals seeking employment opportunities in more prosperous countries but cannot migrate to developed nations. For example, residents of smaller, economically challenged islands may migrate to larger islands with stronger economies, such as moving from Eastern Caribbean countries to

Barbados or Trinidad and Tobago (Thomas-Hope, 2022). The migration of individuals from larger, more economically developed countries to smaller island states is a relatively recent phenomenon.

Trinidad and Tobago is located just 12 miles northwest of Venezuela and has an estimated population of 1.4 million (Central Intelligence Agency, 2024). The racial and ethnic composition includes Indo-Caribbean (35.4%), African-Caribbean (34.2%), bi-racial (African and Indo-Caribbean) at 7.7%, and multiracial and other ethnicities including Syrian, Lebanese, Chinese, indigenous peoples, and European at 12.8% (Central Intelligence Agency, 2024). In 2020, there were a recorded 78,849 immigrants entering Trinidad and Tobago with a large representation from Venezuela (Thomas-Hope, 2022). Currently, Trinidad and Tobago is also one of the largest refugee-receiving countries in the Caribbean (Thomas-Hope, 2022).

In recent years, there has been a significant influx of Venezuelans migrating to neighboring countries, driven by the economic and humanitarian crisis in Venezuela. Hyperinflation and shortages of food, medicine, and healthcare have forced many to seek refugee status in other countries (Huggins & Modeste-James, 2023). Countries like Trinidad and Tobago, Guyana, Colombia, and several Caribbean islands have experienced an increase in Venezuelan migrants. It is estimated that more than 6.1 million Venezuelans have fled the country seeking asylum in other countries (United Nations High Commissioner for Refugees [UNHCR], 2020) and over 60,000 Venezuelans have migrated to Caribbean states such as Trinidad and Tobago. These figures may underestimate how many enter the country through clandestine channels (i.e., sea route) daily (Modeste-James & Huggins, 2022). Venezuelans are not given legal status easily in Trinidad. They generally struggle with being in detention, refoulement back to Venezuela, and face problems with gaining full legal status and safe working conditions in Trinidad, increasing their vulnerability to dangerous situations like illegal trafficking and discrimination (Thomas-Hope, 2022).

In response to the growing number of asylum seekers, the Trinidad and Tobago Government adopted a Refugee Policy in 2014 and UNHCR established an office in January 2016 (UNHCR, 2017, p 1.). Although it is important to recognize the needs of migrants, it should not be at the cost of the citizens in the country. Caribbean governments and regional organizations continually work to address these challenges and improve their social welfare systems with policy reforms, seek international cooperation, and innovative solutions to provide better social services and support for their populations. The Venezuelan refugee crisis has strained Trinidad and Tobago by burdening the public and private sectors with increased need of services, causing physical overcrowding, and civil unrest between its citizens and migrants. This situation underlines the heightened social vulnerability in Trinidad and Tobago, where the pressures from the Venezuelan refugee crisis exacerbate existing tensions and resource constraints, thereby deepening the challenges faced by both public and private sectors.

Social Vulnerability to Climate Change

Social vulnerability is inherently dynamic, influenced by a spectrum of factors, including government policies and investments, prevailing stressors, and emerging opportunities (Tucker et al., 2013). Social vulnerability by country evaluates its economic fragility and resilience against external pressures like climate change, measuring a system's susceptibility to deterioration and its ability to protect itself or recover from stresses and shocks from external forces (Tucker et al., 2013). One such indicator is social welfare systems as it is constantly exposed to processes of change which can be an improvement or degradation of its systems. Social welfare systems (e.g., healthcare, education, housing) are established by governments or other institutions (e.g., Non-governmental agencies) to provide essential support and services to individuals and communities, particularly those who are vulnerable or disadvantaged. It also shapes the cumulative impacts of people's daily actions, such as migration and economic diversification (Tucker et al., 2013). Notably, even within the same geographic location, different groups or households may experience vastly distinct levels of vulnerability.

Trinidad and Tobago Poverty Rate

Recognized as a high-income Caribbean nation, Trinidad and Tobago's economy is based on hydrocarbons exportation, tourism, and a financial services hub (Central Intelligence Agency, 2024). However, the country faces challenges such as high inflation and a growing public debt. Utilizing the multidimensional poverty indicator which assesses 10 indicators covering health, education, and standard of living (The Institute for

Gender and Development Studies, 2015) in 2021, 20% of the population live in poverty and are categorized as multidimensionally poor, with an additional 3.7% (roughly 56,000 people) vulnerable to such poverty and 38% are affected by some level of deprivation which is exacerbated by high inflation prices and increasing debt (Central Intelligence Agency, 2024). The overall unemployment rate is 3.85% (this may not include immigrants; Central Intelligence Agency, 2024), with youth unemployment (ages 15–24) at 12.7% (Central Intelligence Agency, 2024). These disproportionate rates of poverty, unemployment, and depravity make the dependency ratio (e.g., economic reliance on government) relatively high at 43.7% with specific ratios for youth (ages 0–14) at 29.3% and the elderly (ages 65 +) at 16.8% (Central Intelligence Agency, 2024). These statistics may not include undocumented immigrants as they are not counted in the population.

With regard to Venezuelan nationals who are registered and authorized to work, most are unskilled laborers and therefore may not contribute to the development of the country depending on the employment opportunities (Thomas-Hope, 2022; UNHCR, 2021). Venezuelans depend on the United Nations High Commissioner for Refugees (UNHCR), and the Government of Trinidad and Tobago partnered with the local non-governmental organization, Living Waters Community (LWC; Thomas-Hope, 2022). However, the Trinidad and Tobago government pays for access to healthcare and their public safety (Huggins & Modeste-James, 2023). With these staggering statistics, Trinidad and Tobago is at critical risk of catastrophic impact in the event of a severe climate change-related disaster.

Climate Change Vulnerability

Social vulnerability to climate change in Trinidad and Tobago can be divided into three key dimensions, each playing a critical role in shaping the overall resilience of this country. One dimension is exposure, as one facet of vulnerability pertains to the extent of an area's exposure to climate change-related hazards (Tucker et al., 2013). Although Trinidad is geographically outside the usual path of hurricanes and other tropical storms (Central Intelligence Agency, 2024), it is susceptible to earthquakes, rising sea-levels and droughts. The last substantial earthquake was in 2018 with a 6.9 magnitude. Although there were limited causalities, it caused a lot of structural damage to homes, automobiles, the roads, and highways, as it cost the government millions in restoration (Daily Express, 2018).

Another area of concern is water shortages, although Trinidad and Tobago have not registered a drought in years, for many parts of the country there has been excessive water shortages. For the months of December through February 2024, the dryness levels across the country were observed as borderline dry to very dry (Lall, 2024). Droughts present complex challenges that impact employment, health, and the environment (Lall, 2024). The reduced availability of water results in crop failures, diminishing agricultural productivity and leading to financial losses for farmers due to damaged or unharvestable crops (Lall, 2024). There are also supply chain disruptions for businesses that rely on water-intensive processes face disruptions due to limited water availability.

An additional significant public health concern in Trinidad is the exposure to Saharan dust. This dust originates from the Saharan and Sahel desert regions in Africa, consisting of carbonate, oxides, microorganisms, and particulate matter (Akpinar-Elci et al, 2015). In Africa, this dust is exacerbated by intensified land use (e.g., farming, industrialization) and changing climate conditions (e.g., droughts) and it is carried across the Atlantic Ocean to the Caribbean by trade winds. Saharan dust reaches the Caribbean primarily from March through November, spanning both the rainy (e.g., frequent rainfalls) and dry (e.g., no rain falls) seasons (Akpinar-Elci et al, 2015). Often, the dust levels surpass the air quality standards set by the World Health Organization (Akpinar-Elci et al, 2015). Inhalation of these fine particles can exacerbate asthma symptoms and lead to increased hospitalizations (Akpinar-Elci et al, 2015). When combined with the region's high heat temperatures, this dust can significantly affect public health, potentially reducing life expectancy and productivity (Akpinar-Elci et al, 2015). Given these varied climate-related challenges plaguing Trinidad and Tobago from earthquakes and water shortages to the impact of Saharan dust, it is clear more adaptive measures are needed to enhance its resilience and safeguard the health and well-being of its population.

An additional factor contributing to social vulnerability is the sensitivity of a system or community to the impacts of climate change, which can be either adverse or beneficial (Tucker et al., 2013). This sensitivity is influenced by several factors, including the level of dependence on natural resources, the age distribution and health status of the population, and the availability of alternative livelihoods (Tucker et al., 2013). In Trinidad,

34% of the Gross Domestic Product (GDP) is derived from oil and natural gas production, making it the largest producer of these resources in the Caribbean (Central Intelligence Agency, 2024). The extraction of natural gas has significant impacts on climate change as methane is a greenhouse gas and it is a primary component of natural gas. Additionally, the hydraulic fracturing process, or fracking, used to extract natural gas, is energy-intensive and can lead to environmental degradation, including land and water resource depletion.

However, natural gas production has been declining due to competitive global markets. Trinidad and Tobago's government has done little to diversify its economy beyond oil and gas which has been depleting (Reddy, 2017). With the economy experiencing significant fluctuations such as shrinking by 9.08% in 2020 and 1.04% in 2021 due to the COVID-19 pandemic, before a modest recovery of 1.48% in 2022, it is evident that Trinidad's economy is highly sensitive to global market dynamics affecting its natural resources and climate change (Central Intelligence Agency, 2024).

Health Disparities

Trinidad and Tobago's demographic age distribution in 2023 was 19.04% consisting of 0–14 years (136,733 males and 131,316 females), 67.4% are between 15 and 64 years old (482,427 males and 466,200 females), and 13.56% are 65 years and over (88,531 males and 102,253 females; Central Intelligence Agency, 2024). This does not include the undocumented migrants. The life expectancy is 76 years for both genders with women living five years longer than men. Migrants have a decreased life expectancy and worse overall health compared to the host country (Huggins & Modeste-James, 2023). Despite Trinidad and Tobago relatively large working-age population, there is also a high prevalence of non-communicable diseases (e.g., diabetes, hypertension) and cancer (Pan American Health Organization, 2021) as well as, one of the highest HIV/AIDS prevalence rates outside of sub-Saharan Africa (PANCAP 2020; UNAIDS 2022) which makes some incapable of keeping employment. Migrants experience a higher prevalence of mental health disorders and exacerbated physical ailments due to the precipitating factors that led them to migrate from their origin country as well as the effects of the migratory experience (Huggins & Modeste-James, 2023).

The healthcare system, although it is low cost or free, is often inefficient due to long waiting time, inadequate health financing, poor maintenance, and social in-equality with migrants as the last priority (Berkeley, 2019). For migrants, access to healthcare is limited to emergency services which include stabilization of an acute illness and antenatal care, prevention and follow-up care is not permitted (Huggins & Modeste-James, 2023). The role of social work in addressing healthcare challenges, particularly in systems that are inefficient and inequitable, is multifaceted and at times ineffective. Social workers should provide emotional and practical support to individuals navigating these challenges, helping them manage the complexities of healthcare systems while advocating for broader social justice and equity in healthcare delivery. Since most social workers are English-speaking and the migrant patients are primarily Spanish-speaking, it is difficult to develop a working relationship with the lack of translation services. Social workers should work to bridge gaps in care by providing resources, support, and guidance to individuals who may be marginalized by the system. Most social workers are not aware of the resources that migrants can access since most NGOs do not advertise their services publicly, which leaves the support a social worker can provide limited. To address these demographic and health challenges effectively, Trinidad and Tobago must implement comprehensive healthcare reforms that enhance service delivery, improve efficiency, and ensure equitable access for all segments of the population, including vulnerable migrant communities.

Poverty

There is also a marked disparity in income levels, with a significant portion of the population grappling with poverty and a lack of social protections to shield them from economic hardships. According to The World Bank (2022), about 20% of the population lives in poverty, and 38% lack at least one quality of life indicator. Contributory to this situation is the gender wage gap, where women earn 35.3% less than men, often because they are disproportionately employed in lower-paying jobs (Reddy, 2017). Additionally, 75% of single-parent families are headed by women, who not only care for but also financially support their children (Reddy, 2017). These single-parent, female-headed households are six times more likely to live below the poverty line, highlighting a critical area of social vulnerability (Reddy, 2017). For the undocumented immigrants who are working off the books and earning less than minimum wage they are also living below the poverty line.

Climate change also impacts those whose livelihoods and employment depend on climate-sensitive sectors such as those living in rural communities and diminishes capacity for resilience with regard to shocks and stresses (Shah et al., 2013). Individuals involved in tourism, agriculture, or the informal sector (such as street vendors and market sellers), often operate independently as solopreneurs are closely tied to climatic conditions, making them highly vulnerable to climate-induced changes (Shah et al, 2013). Many immigrants are employed in informal jobs, often earning below minimum wage due to their off-the-books status, which leaves them vulnerable to inconsistent income. A sustainable livelihood is one that can withstand and recover from disruptions, sustain, and potentially enhance its capabilities and assets, all while conserving the natural resource base (Bowen, 2006). Addressing these issues demands a comprehensive policy approach that not only targets poverty alleviation and economic inequality, but also integrates climate resilience into all sectors, ensuring that vulnerable communities can sustain and enhance their livelihoods in the face of environmental challenges.

Crime

Another pressing issue is the notably high crime rates with a prevalence of serious offenses such as homicides, robberies, and gang-related activities (Bissessar, 2014). The Caribbean's unique geography and strategic location, sandwiched between South and North America have made it a significant focal point for both drug and human trafficking activities in the broader context of transnational organized crime (Bissessar, 2014). Trinidad and Tobago frequently serve as both a transshipment point and a place of origin for human and drug trafficking operations, facilitating the movement of narcotics between South and North America and Europe (Bissessar, 2014). Many Venezuelans are vulnerable to human trafficking which involves two forms of exploitation: sexual exploitation, which victimizes women and children forced into commercial sexual activities, and labor exploitation, which ensnares individuals in various forms of forced labor, such as domestic servitude and agriculture (United Nations Office of Drugs and Crime, 2023). Given these complex and intertwined issues, Trinidad and Tobago is sensitive to the impact of climate change as they are handling significant and complex social issues with their citizens as well as migrants.

The last social vulnerability factor is adaptive capacity, which entails the ability of natural systems and communities to adapt and respond to the challenges posed by climate change. It encompasses a spectrum of factors, including institutional arrangements, government policies and regulations, and social welfare and entitlement security measures designed to safeguard vulnerable populations. The colonial legacy has left profound and enduring marks on Caribbean governments, often manifesting culturally in political systems inherited from colonial governance, which are frequently characterized by corruption, political instability, and economic inequality. There is little incentive to be adaptive because of organizational structure. This lack of incentive for adaptation is further exacerbated by a bureaucratic inertia, which often slows down or even impedes the implementation of necessary reforms and adaptation strategies, leaving communities at risk and unprepared for climatic impacts.

Social Welfare

Historically social welfare in developing countries that were previously colonized was initially introduced to serve the colonial powers' needs, focusing on social control and acculturation rather than aiding local development (Cox & Pawar, 2013). Modern welfare systems encompass health, education, and law and order, crucial for societies facing challenges like unemployment, high crime rates, and inefficient governmental infrastructures, which impair individual and collective functioning (Popple & Leighinger, 2018). Social welfare provides essential mechanisms that allow individuals to function interdependently within society, offering laws, programs, and services that meet basic social needs crucial for well-being and societal stability (Wickenden, 1960). Overall, it supports vulnerable populations, advocates for marginalized groups, and ensures aid during disasters, reflecting its role as a vital organized function in adapting to changing social conditions (United Nations, 1967).

The Residual Welfare Model

These services are usually disseminated within three distinct classifications, each offering a unique perspective on the role of social welfare services. The first classification is the residual welfare model, which asserts that social welfare services should only be extended when an individual's needs are inadequately met (Titmuss, 1974, 1987). This model views the proper function of a modern industrial society as aiding individuals in achieving self-fulfillment, making services accessible to all, not just those living in poverty. Under the residual welfare model, social welfare is considered a last resort or "residual" measure, provided only when needs cannot be met through employment or family support. In this framework, government involvement in welfare is minimal, often leaving the provision of basic necessities such as food, shelter, and emergency assistance to the private sector, charities, and community organizations. Assistance is targeted at those in extreme poverty or significant crisis, and there is often a stigma associated with receiving benefits, as it is perceived as a measure for those in dire circumstances. Social workers in this model act as advocates for those in poverty or significant distress, helping them navigate the limited social services available. They connect individuals with the resources necessary to meet immediate needs, while also addressing the stigma associated with receiving aid by providing emotional support and advocating for their clients' dignity and rights. However, since government involvement is minimal in the residual model, social workers must often rely on non-governmental resources and work within a limited framework of support, which can limit their ability to address the underlying structural issues related to poverty and inequality. Midgley (2012) has argued that this model fails to provide adequate support for vulnerable populations and does not address the structural issues underlying poverty and inequality.

Industrial Achievement/Performance

The second classification, known as industrial achievement/performance, also known as the "handmaiden model" emphasizes the development of specific social welfare programs that have a positive impact on the economy (Titmuss, 1974). In this context, welfare is seen as a catalyst for economic growth, enhancing the workforce's capacity and contributing to overall economic prosperity. This model suggests that well-designed social welfare initiatives can contribute to economic growth and improved performance by enhancing the capacity and productivity of the labor force (Titmuss, 1974). It posits that providing support and resources to individuals, such as education and training, healthcare, and childcare, can lead to a more skilled and productive workforce, which leads to economic growth and competitiveness. It may include programs and policies aimed at vocational training, job placement, and lifelong learning opportunities to enhance employability (Titmuss, 1974). In this model, social workers operate within a framework that aligns social welfare with economic goals, focusing on enhancing individual productivity and employability. In this context, social workers act as facilitators who help individuals access services such as education, vocational training, healthcare, and childcare, which are designed to improve their ability to participate in the workforce. Collaboration between the public and private sectors is often seen as essential in the industrial achievement/performance model (Titmuss, 1974). Governments may fund and work closely with businesses and industry stakeholders to align social welfare policies with economic objectives. This model prioritizes economic outcomes and the perspectives of industry over social justice, equity, and needs of the individual in society.

Institutional Redistributive Model of Social Policy

Lastly, the third classification is the Institutional Redistributive Model of Social Policy. This conceptual model integrates institutions in society, providing universalist services outside the market on the principle of need (Titmuss, 1974). It concentrates on the multiple effects of social change and the economic system, and in part on the principle of social equality. It is basically a model incorporating systems of redistribution in command-over resources-through-time. This model places a strong emphasis on the government's role in actively redistributing resources and opportunities to ensure a more equitable distribution of wealth and income (Titmuss, 1974). Social workers in this model often focus on addressing systemic inequalities by working at both individual and policy levels. At the individual level, they help people navigate public services and ensure that their rights to access universal benefits are upheld. They advocate for clients facing barriers in accessing services, ensuring marginalized populations receive the support they need. This framework is designed to implement social welfare policies with a primary focus on reducing income and wealth inequalities within society. The model often advocates for the provision of universal social services, such as healthcare, education, and childcare, to ensure that all members of society have equal access to essential services regardless of their income or social status (Titmuss, 1974). Critiques of the Institutional Redistributive Model of

Social Policy often centers around concerns related to heavy taxation and wealth redistribution measures which can stifle economic growth and hinder economic efficiency (Titmuss, 1974). The model's generous social safety nets and income redistribution can create disincentives for individuals to seek employment or strive for higher incomes. It may inadvertently foster dependency on government support, discouraging individuals from taking responsibility for their own economic well-being and self-sufficiency (Titmuss, 1974). The one-size-fits-all approach of universal social services may not address the diverse needs of individuals and communities. This model can lead to inefficient resource allocation and fail to adequately address specific challenges faced by different groups.

Trinidad and Tobago's welfare system is a multifaceted blend of various conceptual frameworks, primarily drawing from the Residual, Industrial Achievement/Performance, and Institutional Redistributive Models of Social Policy. This unique combination reflects the country's approach to ensuring the well-being of its citizens. The government applies the Institutional Redistributive Model in promoting equal distribution of services and implementing measures to foster a more equitable society. In this context, the country offers free healthcare and education to all of its citizens, regardless of their income. Furthermore, Trinidad and Tobago's welfare system also incorporates elements of the Industrial Achievement/Performance Model. This framework emphasizes the role of education in contributing to economic growth and enhancing the overall performance of the labor force. The country provides free education from primary school through higher education. Notably, Trinidad and Tobago implemented the Government Assistance for Tuition Expenses (GATE) program, benefiting approximately 200,000 national students pursuing education at colleges and vocational institutions (Simms, 2016). In return for this support, graduates are expected to contribute to the country's workforce for a specified period, which varies based on the value of their grant. This approach underscores the nation's commitment to nurturing its citizens through education, with the aim of improving their capabilities and productivity for the country.

While education in Trinidad and Tobago is provided free of charge, it has historically been restricted to citizens, with migrant children unable to access the national school system. However, the recent decision by the Government of Trinidad and Tobago to include refugee and migrant children in the primary school system starting from July 2023 is a significant and commendable step forward. This decision has been welcomed by the United Nations Resident Coordinator and members of the UN Country Team in Trinidad and Tobago. They recognize that access to education is a fundamental right that profoundly impacts the lives of children, their families, and even the future of the host country. Sending children to school not only facilitates family progress but also aids in integration, language acquisition, and provides a sense of normalcy and security. Education opens doors to envisioning brighter futures and finding pathways to economic and social stability for families.

Trinidad and Tobago's welfare system also follows a residual model, particularly when it comes to catering to the elderly, individuals in need, and the disabled. In this approach, government involvement in the provision of social welfare services is minimal, and the responsibility largely falls upon the private sector, charitable organizations, and community groups to address social needs. The residual model typically encompasses a limited scope of welfare programs, primarily concentrating on essentials such as food, shelter, and emergency assistance. Trinidad and Tobago with the assistance of United Nations High Commissioner for Refugees (UNHCR), are trying to manage the current wave of migrants while trying to manage competing social problems. In light of the limited scope and reliance on non-governmental actors inherent in Trinidad and Tobago's residual welfare model, the collaboration with organizations like the UNHCR is crucial as the country navigates the dual challenges of supporting vulnerable populations and managing a significant influx of migrants.

Overall, social workers in Trinidad and Tobago's welfare system function as connectors between individuals and institutions, advocating for equitable access to resources, promoting self-sufficiency, and supporting individuals as they navigate systemic challenges. In Trinidad and Tobago's multifaceted welfare system, social workers should play a critical role in addressing the diverse needs of the population, particularly within the frameworks of the Residual, Industrial Achievement/Performance, and Institutional Redistributive Models of Social Policy.

Within the Institutional Redistributive Model, social workers should advocate for universal access to essential services like healthcare and education, ensuring that all citizens and migrants, regardless of income, can benefit. In the Industrial Achievement/Performance Model, social workers focus on enhancing the capacity of individuals to participate in the workforce and contribute to the country's economic growth by connecting them to educational programs like the Government Assistance for Tuition Expenses (GATE). By supporting education and vocational training efforts, social workers play a vital role in fostering personal and societal development. In the Residual Welfare Model, social workers respond to immediate needs for food, shelter, and emergency assistance, particularly for those in crisis. This model often applies to the migrants, who may rely on minimal government support. Social workers collaborate with Non-Governmental Organizations (NGOs), like the UNHCR, to bridge gaps in services and ensure that vulnerable groups receive essential care and resources. Social workers act as intermediaries between migrants and these external organizations, helping them secure necessities such as food, shelter, healthcare, and legal assistance. They also play a critical role in helping migrants understand their rights, especially regarding access to emergency services.

Conclusion

The social welfare system in Trinidad and Tobago has significant implications for migrants, particularly given its historical and structural foundations. Historically, social welfare systems in Trinidad and Tobago were designed to serve colonial powers by enforcing social control and acculturation rather than fostering local development. Today, modern social welfare systems are essential for addressing key social challenges, such as displacement and crime, and providing crucial services like healthcare and education, but it is rife with disparities as who can access services and who cannot.

Trinidad and Tobago's use of the residual welfare model where government intervention is minimal and support is often left to charities and NGOs means that many migrants face limited access to necessities such as food, shelter, and emergency aid. This model tends to focus on crisis intervention, rather than providing ongoing support, which does not foster integration and the advancement of migrants. Prior to the influx of Venezuelans to Trinidad and Tobago, migrant children were excluded from the school system. The recent efforts to include them mark a positive step toward integration. However, despite these advancements, the broader system continues to rely heavily on international organizations like the UNHCR to provide services for migrants. This reliance points to the limitations of the residual welfare model in addressing the long-term needs of both citizens and migrants, as it prioritizes crisis responses over systemic solutions. The ineffectiveness of social workers is rooted in the limitations of the residual welfare model, which provides only minimal government intervention and leaves much of the responsibility for migrant support to charities and NGOs. Social workers are forced to operate in a reactive capacity, offering emergency aid such as food, shelter, and temporary relief, rather than long-term solutions that foster integration, self-sufficiency, and the advancement of migrants.

Climate change, with its effects such as rising sea levels, more intense hurricanes, and prolonged droughts, disproportionately affects small island developing states like Trinidad and Tobago. Climate change has the potential to deepen inequalities, making migrants more susceptible to the negative impacts of environmental disasters. Migrants arriving in Trinidad and Tobago not only face the immediate challenges of resettlement but also contend with the effects of environmental degradation within the host country which affect housing, healthcare, and employment opportunities. For instance, migrants who are employed in informal sectors or live in coastal areas in illegal housing structures are particularly vulnerable to extreme weather events. The residual welfare model's crisis-oriented approach means that support is typically reactive, leaving both citizens and migrants exposed to the long-term challenges associated with climate change. Social workers operate on the front lines during these environmental crises, providing short-term support to displaced or affected migrants. However, the residual model's focus on addressing immediate needs rather than systemic solutions limits their ability to foster long-term resilience. Instead of being able to create sustainable pathways for migrants to secure safe housing, stable employment, or healthcare, social workers are often left managing the aftermath of climate-related disasters without the necessary tools or resources to implement proactive strategies.

As climate change continues to drive displacement, a more comprehensive and adaptive welfare system will be necessary to support both the local population and migrants. This system must focus on long-term sustainability, resilience-building, and equitable access to resources, ensuring that all individuals, regardless of their migration status. Trinidad and Tobago's preparedness for climate change involves a multifaceted approach, considering the dimension of exposure to climate-related hazards, sensitivity to these impacts, and the capacity to adapt and respond effectively (Tucker et al., 2013). Social workers working in the social welfare system must advocate for social equity, to ensure that vulnerable populations have access to the resources they need and address the systemic issues that exacerbate social and environmental vulnerabilities. In terms of exposure, the country acknowledges the importance of assessing the extent of its vulnerability to climate change-related hazards. This includes evaluating the likelihood and intensity of events such as earthquakes, sea-level rise, and prolonged droughts, which have the potential to significantly impact specific locations and the health of its population within Trinidad and Tobago.

Trinidad and Tobago should recognize the degree of susceptibility to climate change impacts varies across systems and communities. Factors influencing sensitivity span a wide spectrum, encompassing the level of dependence on natural resources, the age distribution and health status of the population, and the accessibility of alternative livelihoods. By understanding these factors, the country should tailor its strategies and interventions to address the unique vulnerabilities of different communities and sectors. This involves the development of policies, strategies, and mechanisms that enhance resilience, promote sustainable practices, and empower communities to manage climate-related risks and uncertainties. Climate change disproportionately affects vulnerable populations, particularly in terms of displacement, loss of livelihoods, food insecurity, and health impacts. Social workers should advocate for climate justice, emphasizing the need for equitable policies that protect marginalized communities from the effects of climate change.

References

Akpinar-Elci, M., Martin, F., Behr, J., & Diaz, R. (2015). Saharan dust, climate variability, and asthma in Grenada, the Caribbean. *International Journal of Biometerology, 59*, 1667–1671.

Anderson, B., Andrijasevic, R., Bales, K., Chan, J., Ciancanelli, P., Bastos, F. B. C. C., & Withers, M. (2022). *Global labor migration: New Directions*. Illinois: University of Illinois Press. https://doi.org/10.5622/illinois/9780252044700.001.0001.

Baldacchino, G. (2018). Seizing history: Development and non-climate change in Small Island developing States. *International Journal of Climate Change Strategic Management, 10*, 217–228.

Barnett, J & Waters, E. (2016). Rethinking the vulnerability of small island states: Climate change and development in the Pacific Islands. In Grugel, J. & Hammett, D. (Eds.), *The Palgrave handbook of international development*. London: Palgrave Macmillan.

Basta, E. (2023). *The impact of emigration on British society since World War II*.

Beckles, H. (2016). *The first black slave society: Britain's Barbarity time in Barbados 1636–1876/* Kingston: The University of the West Indies Press.
Bennie Berkeley (2019). An analysis of the problems users experience in public health care facilities in Trinidad and Tobago (pp.1–13). *Open access library Journal.*
Bissessar, A. (2014). The nexus between structural adjustment and the emergence of gangs: The case of Trinidad and Tobago. In R. Seepersad & A. Bissessar (2014). *Gangs in the Caribbean* (pp. 131–149). Cambridge Scholars Publishing.
Bowen, G. (2006). The challenges of poverty and social welfare in the Caribbean. *International Journal of Social Welfare, 16*, 150–158.
Clarke, J. (2014). The end of the Welfare State? The challenges of deconstruction and reconstruction. *The end of welfare as we know it*, 19–34.
Central Intelligence Agency (2024). The World Factbook: Trinidad and Tobago. cia.gov/the-world-factbook/countries/trinidad-and-tobago/#people-and-society
Cox, D., & Pawar, M. (2013). *International Social Work: Issues, Strategies, and Programs*. Sage Publications.
Daily express (2018). Monster quake terrifies T&T: Damage widespread. Daily express. Monster quake terrifies T&T: Damage widespread | Local News | trinidadexpress.com
Forbes, J. (2014). Caribbean Youth and Unemployment. Caricom. Caribbean Youth and Unemployment – CARICOM
Herbert, C. (2021). FIELD WORK More Venezuelans in Trinidad in the 21st century A brief account and analysis of the first large-scale irregular migration of Venezuelans to Trinidad and Tobago in recent years. *International Journal of Anthropology and Ethnology, 5*, 15.
Huggins, C. (2021). Through the eyes of the perpetrator: the historical and contemporary cultural context of intimate partner violence in the Caribbean. In Bissessar & Huggins (2021) *Gender and Domestic Violence in the Caribbean*. Switzerland: Palgrave.
Huggins, C. & Modeste-James. (2023). Providing healthcare to Venezuelan migrants in the small island states of Trinidad: Physician's lived experience. In Wallace, W. (Eds.), *The movement of Venezuelans to the Americas and the Caribbean in the 21st century.* Palgrave.
Lall, R. (2024). Drought could damage economy. Trinidad and Tobago Guardian. 'Drought could damage economy'—Trinidad Guardian.
Lister, M. (2014). Climate change refugees. *Critical Review of International Social and Political Philosophy, 17*(5), 618–634.
Longépée, E. (2016). La resilience des atolls de Kiribati (Micronésie) à la sécheresse: comparaison entre un système urbain et un système rural. *Cybergeo: European Journal of Geography.*

Kazana-Wiśniowiecki. J. (2023). UN in T&T welcomes decision to include migrant and refugee children in national education. United Nations UN in T&T welcomes decision to include migrant and refugee children in national education —Trinidad and Tobago | ReliefWeb

Keshen & Lazickas, (2022) Migration and migrants: A global overview. International organization for migration.

Midgley, J. (2012). *Social development: Theory and practice.* Sage Publications.

Modeste-James, A. & Huggins, C. (2022). The impact of domestic violence on children in the Caribbean. In Bissessar, A. M. & Huggins, C. (Eds.), *Domestic violence in the anglophone Caribbean*, 1–12. Palgrave Macmillan.

Nurse, L. A., McLean, R. F., Agard, J., Briguglio, L. P., Duvat-Magnan, V., Pelesikoti, N., Emma Tompkins, Webb, A. (2014). Small islands. *Climate change 2014: Impacts, adaptation, and vulnerability. Part B: Regional aspects. Contribution of working group II to the fifth assessment report of the intergovernmental panel on climate change* (pp-1613).

Office of the High Representative for the least developed countries, Landlocked developing countries and Small Island developing states. (n.d.). Small Island developing States (SIDS) are a distinct group of 39 States and 18 Associate Members of United Nations regional commissions that face unique social, economic and environmental vulnerabilities. About Small Island Developing States | Office of the High Representative for the Least Developed Countries, Landlocked Developing Countries and Small Island Developing States

Pan-American Health Organization. (2021) Trinidad and Tobago: Empowering communitiesto prevent and self-manage noncommunicable diseases. Trinidad and Tobago: Empowering communities to prevent and self-manage noncommunicable diseases—PAHO/WHO | Pan American Health Organization

Pan Caribbean Partnership Against HIV/AIDS. (2020). Seizing the moment: tackling entrenched inequaltities to end epidemics. Global AIDS update

Petzold, J., & Magnan, A. (2019). Climate change: Thinking small islands beyond Small Island Developing States. *Climate Change, 152*, 145–165.

Popple, P., & Leighninger, L. (2018). *The policy-based profession: An introduction to social Welfare policy analysis for social workers* (7th ed.). Pearson.

Reddy, A. (2017) Why is Trinidad and Tobago poor? Borgen Report. https://borgenproject.org/why-is-trinidad-and-tobago-poor.

Scobie, M. (2018). Accountability in climate change governance and Caribbean SIDS. *Environmental Development Sustainability, 20*, 769–787.

Shah, K., Dulal, H., Johnson, C., & Baptiste, A. (2013). Understanding livelihood vulnerability to climate change: Applying the livelihood vulnerability index in Trinidad and Tobago. *Geoforum, 47*, 125–137.

Simms, E. (2016). Report of the gate task force. Report of the gate task force (ttparliament.org)

Spencer, N., & Strobl, E. (2020). Hurricanes, climate change, and social welfare: Evidence from the Caribbean. *Climatic Change, 163*, 337–357.

The Institute for Gender and Development Studies, St. Augustine Unit (2015). *Perspectives On Well-Being, Progress and Vulnerability In Trinidad And Tobago Based on Focus Groups and In-depth Interviews conducted within the framework of the 2016 Regional Human Development Report on Multidimensional Progress in Latin America and the Caribbean. United Nations Development Programme.* 6448ed60e3ef4c0daac88c3bbb34ec31591b9fd01f2245bcd95a990405647835 (1).pdf

Thomas-Hope (2022) *Migration in Trinidad and Tobago*: Current trends and policies. UNDP Latin America and the Caribbean policy documents series. Migration in Trinidad and Tobago: current Trends and Policies (undp.org).

Titmuss, R. M. (1963). The welfare state: Images and realities. *Social Service Review, 37*(1), 1–11.

Tucker, J., Daoud, M., Oates, N., Few, R., Conway, D., Mtisi, S., & Matheson, S. (2013). Social vulnerability in three high-poverty climate change hot spots: What does the climate change literature tell us? *Regional Environmental Change, 13*(1), 1–14. https://doi.org/10.1007/s10113-012-0341-2

UNAIDS. (2022). Global AIDS update 2022: In danger. Joint United Nations Programme on HIV/AIDS. https://www.unaids.org.

UNHCR. (2017). UNHCR Global report 2017. United Nations High Commissioner for Refugees. UNHCR Global Report 2017 | UNHCR

UNHCR. (2021). UNHCR Global Report 2021. United Nations High Commissioner for Refugees. Document—UNHCR Global Report 2021

United Nations Development Programme (2023). *Multidimensional poverty index 2023. Unstacking global poverty*: Data for high impact action briefing note for countries on the 2023 Multidimensional poverty index Trinidad and Tobago What is the global Multidimensional Poverty Index?. TTO.pdf (undp.org)

United Nations Office on Drugs & Crime (2023). The Caribbean takes actions to combat human trafficking. The Caribbean takes action to combat human trafficking (unodc.org)

Wickenden, E. (1960). Social security and voluntary social welfare. *ILR Review, 14*(1), 94–106.

Williams, C. & Valle, S. (2023). *Trinidad invites neighbours to process oil and gas in the country*. Reuters. Trinidad invites neighbours to process oil and gas in the country | Reuters

World Bank (2022). *Understanding poverty, global development data*. Understanding poverty (worldbank.org)
World population review. (2023). *Caribbean population*. Accessed November 2023. https://worldpopulationreview.com/countries/trinidad-and-tobago-population/. Caribbean Population 2023 (worldpopulationreview.com)

CHAPTER 10

Mitigation and Adaptation Innovations Addressing Climate Change and Forced Migration: Case Study Research in Five World Regions

Carmen C. Monico, Shalee N. Forney, Helen B. Tadese, and Ashley D. Gonzalez

Introduction

Climate change is one of the grand challenges humans and the entire planet are facing. Because climate change is a major contributor of forced migration, innovative mitigation and adaptation strategies are critically needed. This chapter analyses the global policy environment, offers a definitional framework and discusses the relationship between climate

C. C. Monico (✉) · S. N. Forney
North Carolina Agricultural and Technical State University (NC A&T), Greensboro, NC, USA
e-mail: ccmonico@ncat.edu

H. B. Tadese · A. D. Gonzalez
University of North Carolina Greensboro (UNCG), Greensboro, NC, USA
e-mail: hbtadese@uncg.edu

D. D. Joseph et al. (eds.), *Climate Change*,
https://doi.org/10.1007/978-3-031-83037-2_10

change and climate refugees. The chapter also examines five world regions impacted by climate change and where mitigation and adaptation innovations are developed. Global solutions are reviewed, including the pursuit of climate governance and climate resilience. The chapter ends with an inquiry into the implications of this study.

Literature Review

Global Policy Environment

The 2015 Paris Agreement adopted by 196 parties at the United Nations (UN) Climate Change Conference in Paris, France in 2015, is an international treaty on climate change that has the goal of holding the global average temperature to well below 2°C above pre-industrial levels (United Nations Framework Convention on Climate Change, UNFCCC, 2022). Countries must not only reduce greenhouse gas emissions but they must invest in research on climate change and in the development of creative solutions for mitigation (to reduce greenhouse gas emissions or create renewable energy sources) and adaptation (to prevent or minimize adverse impacts of climate change) (European Environment Agency, 2022; Suarez, 2020). Alternatives in the use of both technology- and nature-based solutions (the use of renewable energy from animal waste, wind and the sun) have been shown to mitigate greenhouse gas emissions (Saidi & Omri, 2020). Climate adaptation requires innovative approaches to adjust to current and future climate change effects (European Environment Agency, 2022). In spite of the international attention to climate change, its nexus to climate migration remains mostly unrecognized; yet, addressing both in tandem is not ranked as a global priority. This chapter focuses on this inherent global paradox in the Grand Challenges for Social Work (2021).

In spite of the UN call to "scale up" global actions (UN, 2021, p.5), the lack of international agreements in preparedness and recovery efforts in response to the humanitarian crisis created by climate change is insufficient and delayed. It was not until 2021 that the United States (U.S.) for the first time acknowledged the impact of climate change on migration, and that migration is necessary for survival and adaptation (White House, 2021). In 2022 participating UN countries in the Climate Change Conference (COP27) went further than past international agreements to establish a mechanism for articulating a global response to the humanitarian crisis emerging from climate change. The unprecedented agreement on establishing a new "loss and damage" fund for vulnerable countries is a step in the direction of alternative solutions to the climate and humanitarian crises resulting from climate disasters (UN, 2022, para. 2) which are adversely impacting countries of origin of many climate refugees. The UN hosted a "climate ambition summit" in 2023, ahead of the "stocktake" of the Paris Agreement, discussed during the COP28 (UN, 2022, para. 24). Despite the gradual onset of climate change, accompanied calamities are devastating and threatening human existence everywhere.

Conceptualizing Climate Change and Climate Refugees

Climate change causes uncertainty in people's lives. Drought and floods affect productivity in the agriculture sector and impact food supply. Low food productivity could lead to a famine often negatively impacts the economy (Thant et al., 2022). On a global scale, climate change affects all planet inhabitants, but vulnerable populations bear the most burden of climate change and associated environmental disasters. Climate change disproportionately affects vulnerable and minority groups in poverty with pre-existing adverse social, political and economic power and representation. Climate change is an environmental justice issue which threatens the physical, mental and social wellbeing of most vulnerable groups.

We (the authors) challenge those using national legal frameworks to determine who deserves protection after forced displacement. Legally based definitions of who deserves protection are exclusionary. Given the growing migration trends, policymakers around the world use these definitions to deny protection rights to migrants, including those who may have been displaced by climate or environmental changes impacting their livelihood. In fact, the Office of the United Nations High Commissioner for Human Rights (United Nations, 2023) has developed a guide to promote a human rights-based approach to migration; that is, to understand "how human rights laws and standards can be operationalized to make migration safer and an empowering experience for all" (United Nations, 2023, p. 2). Although we found historical references and contemporary literature using different terms in distinct ways to refer to those forcibly migrating within and across borders due to changes caused by climate and other environmental changes, we use the term climate refugees to refer to these populations.

The National Aeronautics and Space Administration (NASA) working definition of climate change as "the long-term change in the predictable weather patterns of a place that can be observed in unusually heavy and long rain seasons and extreme heat waves" (NASA, 2023, par. 1). The United Nations Environment Programme (1985) expert, Essam El-Hinnawi, defined environmental refugees as those "having been forced to leave their original habitat (or having left voluntarily) to protect themselves from harm and/or to seek a better quality of life" (p. 4). In this paper, the term climate refugees is used to link it more directly to climate change.

Millions have become internally displaced or forced to emigrate from their countries of origin. The Institute of Economics and Peace (2020) estimated 1.2 billion will be displaced by 2050 globally. The United Nations High Commissioner for Refugees (UNHCR, 2023) estimated that by the end of 2022 there were 108.4 million people experiencing forced displacement due to various causes, including persecution, conflict, violence and human rights violations. Reliable data on people displaced or to be displaced due to climate change is limited (Naser, 2012); however, several global estimates are available. Although conflict has been found a major cause of migration and displacement, numerous studies have linked migration to climate change, which has either contributed or is responsible for the displacement of 7 million people worldwide (UN, 2021). The UNHCR representative Filippo Grandi claimed that "70 % of people displaced within their home countries by conflict and violence, come from countries on the front lines of the climate emergency" (Grandi, 2021, par. 1). The World Bank estimated that threats from hurricanes coupled with the increase in sea levels will displace an estimated two hundred million people worldwide by 2050 if no action is taken to reduce the emission of greenhouse gasses (Riguad, et al., 2018).

Nexus Between Climate Change and Climate Refugee

The United Nations Environment Programme (1985) recognized that the source of displacement is environmental degradation and desertification which destroys their habitat and undermines their livelihood, posing great risks to their health (Apap, 2021). Displaced refugees face adverse impacts in their home countries (before migration), while in transit or in refugee settlements, and at their host countries (destination)

(Shultz et al., 2018). In their home countries climate refugees experience poverty, hunger and political instability. Climate change creates a chain of problems, such as floods that destroy farmland and crops, which may cause hunger and income, which may lead to other social dilemmas such as poverty and violence (Crate & Nuttall, 2016). In transit, climate refugees may face violence, starvation and lack of medical care. Loss of life is a common phenomenon during migration (McMahon & Sigona, 2021); these conditions are often experienced in destination. Climate refugees face discrimination, structural and systemic oppression and possess limited resources to cope with climate disasters (Environmental Protection Agency, 2022).

Advances in science and technology enable the recording of staggering increases in global temperatures and unstable weather conditions. Climate change indicators show more frequent occurrences of wildfires with longer burning cycles (Environmental Protection Agency, 2023). Severe and frequent environmental disasters have become chronic with devastating consequences to populations. Over the last century, environmental calamities such as wildfires, hurricanes and floods have increased (Smith, 2023). Poor infrastructure and limited resources in regions where vulnerable populations reside make recovery efforts difficult to access (Barbier & Hochard, 2018). Climate-induced displacement is affecting the working relationship they have with each other and the way of life for many people across borders (Alam, 2022).

As climate change disasters continue to increase in frequency and severity, so is the threat to the physical, mental and social wellbeing of most vulnerable groups. Drastic increases in weather extremes in high temperatures, longer periods of droughts, increase in forest fires and reduction of glacier ice are climate-related changes that have adversely affected people globally (United Nations Climate Action, 2023). Impoverished regions face severe risk of climate disasters and lack necessary adaptation and mitigation resources (Conway, 2011). Millions from these regions have become internally displaced or forced to emigrate from their countries of origin. Yet, everyone desires to find alternatives where all parties can survive and have a quality of life worth living (Alam, 2022).

Methods

A set of questions guided the case study research, which was the selected method. The questions were: How is climate change impacting human livelihoods and leading to forced migration within and across borders? What are key innovative strategies for mitigation and adaptation within at the national, regional and global levels? Case study research was used to carry out this study on climate change and people who have been forcibly displaced, including internally displaced peoples, immigrants, refugees, asylum seekers and others seeking protection from environmental disaster caused by climate change.

Rigour in case study research can take several forms, depending on the perspective or strategy adopted (Gibbert & Ruigrok, 2010). Its central aim is to gain understanding of the people involved in the research since this approach is an "in-depth, multi-faceted explorations of complex issues in their real-life settings" (Crowe, et. al, 2011, p.11). An element of rigour was determining inclusion and exclusion criteria to select cases. Several world regions were covered, as well as mitigation and adaptation strategies, and innovative approaches to climate change that did not conform with conventional approaches. Excluded were cases that focused solely on non-human species or who focused solely on mitigation or adaptation of an ecological environment. Selection of cases was based on the emphasis on climate change impact on human impact at their origins, transits or destinations.

For the literature review for the case research, we mostly relied on database searches, such as PubMed, Agricultural & Environmental Science, Global Environmental Justice Collection and Google Scholar. We also conducted searches of grey literature among organizations addressing the welfare of climate refugees. Keywords were used when identifying sources.

Results Case Research

A standard format is used for case study reporting. Every case study has an introduction providing some regional background on climate change. It includes a discussion of the main challenges the region is facing with respect to climate change and climate refugees. Finally, every case study examines innovative mitigation and adaptation strategies.

Case 1: Central and South America

Introduction—In recent decades, environmental degradation and drastic climate changes in the countries of origin of South and Central Americans have become the main cause of increased climate migration. This case study exemplifies why people have been migrating within and outside their borders in search of a haven. Communities that rely on their immediate environment for their livelihood cannot afford the drastic changes experienced globally and impacting them locally. It is predicted that "by 2030, climate change could push 3 million people a year in extreme poverty" (Coda, 2021). As people migrate, they tend to go to cities and countries where their families and communities have historically migrated; migration is a way of reuniting families and preserving their distinct cultures. As in the case of regional migration, survival is a great motivator and people still attempt the journey to more prosperous countries. Climate migrants are having a drastic effect on nations as they search for a sustainable life.

Challenges—Climate migration is expected to increase in Central and South America, and countries that have been receiving an influx of migrants are not prepared for the future. This will affect the economies of various countries, as well as their social development. Central America and Mexico are projected to see an increase in climate migrants due to the decrease in water availability and crop productivity in South and Central American countries (Riguad et al., 2018) Colombia has more Venezuelan climate migrants now but is not prepared to take on the millions of migrants to come. Caravans seeking refugee increased in 2018 and have influenced politics in receiving nations, making it difficult for migrants who escape to settle down in new countries (Marchand, 2021).

Mitigation and Adaptations—The World Bank conducted mitigation and adaptation programmes to improve resiliency in Central and South America. The mitigation programme they developed involved several countries in Latin America investing in reducing emissions from forests; four countries (Chile, Costa Rica, Guatemala and the Dominican Republic) have developed unique "jurisdictional results-based agreements for emission reductions from forests" (Coda, 2021, par. 13). This incentive encourages companies to join in sustainable, eco-friendly efforts while also creating a better environment for those who rely on the forests for their livelihoods (Coda, 2021). Yet, impoverished regions of

South America face severe risks of climate disasters and lack the necessary adaptation and mitigation resources (Conway, 2011). Country-level and regional "policy decisions made today will shape the extent to which the effects of climate change will be positive for migrants and their families" (Riguad et al., 2018, p. 181), particularly those displaced by climate change. As a recipient country, the U.S. must develop programmes that make the integration easier in countries of reception and enable better reintegration in countries of origin of deportees (Monico & Pitts, 2021).

Case 2: South Asia

Introduction—Asia contributes to climate change in an impactful way by producing half of the world's carbon dioxide emissions. Five of the biggest greenhouse gas-emitting countries, China, Indonesia, South Korea, Saudi Arabia and India are in Asia (International Monetary Fund, (2020). Livelihoods are sensitive to climate as any drastic changes impact the availability, access and utilization of food. Food scarcity due to climate change leads to food insecurity, which in turn affects people's health (Krishnamurthy et al., 2015).

Challenges—Climate change is drastically affecting rain-fed food production. Over 60% of agriculture nationwide is dependent on rainfall in Asia. Diminished rainfall affects crops and decreases production. Decreased food production impacts food prices, which significantly increase, making it much more difficult for families to access food. Limited water supply and quality also affect local bean production, which is essential for household nutrition. Climate change affects dietary diversity and reduces food consumption leading to disease and health-related problems. These dynamics impact households who are already struggling to afford food and are already living in poverty, especially those residing in the mountainous regions of Nepal (Krishnamurthy et al., 2015).

Mitigation and Adaptations—As part of the South Asian Association for Regional Cooperation (SAARC), disaster risk reduction and climate change are being addressed using a more holistic approach. Afghanistan, Bangladesh, Bhutan, India, The Maldives, Nepal, Pakistan and Sri Lanka implement the Sendai Framework for Disaster and Risk Reduction 2015–2030, which aims at reducing and preventing disaster risk through preparedness, recovery and resilience (Mall et al., 2018). As a result of these efforts, these countries have introduced a National Adaptation Programme of Action (NAPA), which aims to reduce carbon emissions by promoting renewable energy, lessen the impact of flooding and cyclones through the construction of seawalls and revive agriculture by enhancing water availability. Sri Lanka's National Climate Change Adaptation Strategy (NCCAS) promotes the use of biofuels as a substitute for carbon emissions in order to preserve biodiversity and protect livelihood, as well as climate-resilient infrastructure development. India had set a goal to generate 175 gigawatts of renewable energy by 2022 and managed to achieve to date. This has also resulted in the creation of new employment opportunities. These mitigating strategies have created sustainable living conditions for those involved in this programme (Giri et al., 2023).

Case 3: The Mediterranean

Introduction—Surrounded by three continents (Europe, Africa and Asia) the Mediterranean is rich with biodiversity. The effects of climate change in the Mediterranean are become more obvious as "regional warming will continue faster than the global average and will exceed the global mean by 20% on an annual basin and 50% in summer" (Cherif et al., 2020, p. 78). This will affect the health and livelihoods of thousands, with the environment around them slowly deteriorating; people will begin to migrate to other countries searching for a sustainable life, especially as food insecurity rises due to climate change (Cherif et al., 2020). The movement of climate migrants will likely impact the continents bordering the Mediterranean. With the potential of an entire region to have its population flee to countries unprepared for the influx of newcomers, measures taken must aim to deter the adverse climate effects.

Challenges—The warming of the Mediterranean will directly impact the clean water of the region and as the sea level rises residents will have to leave their homes. The coasts are at a greater risk that include flooding, erosion and droughts, which all pose a threat to food security (Cherif et al., 2020). Millions of people will not be able to access basic needs and will seek to meet those needs elsewhere. Climate change has affected the ecosystem which in turn is shrinking the unique biodiversity of the Mediterranean. The world does not have to wait for the effects of climate change to become visible in the region as there are signs of change already. Air pollution in the Mediterranean is exacerbated by sea pollution with plastic, by atmospheric radionuclides that are "deposited from the air onto the land and sea surface by wet and dry deposition" (Cherif et al., 2020, p. 97). In the long run, these dynamics will lead to poor health among the population and may trigger societal divide as people attempt to access cleaner water and air.

Mitigation and Adaptations—Adaptation projects aim at increasing the resilience of the coasts, developing a pilot transnational severe weather service for the Balkan-Mediterranean, and early detection of wildfires (Climate Adapt, n.d.). Several areas of the Mediterranean geographically overlap with European mitigation strategies. Mitigation projects developed include monitoring protocols used in observing marine protected areas (Mediterranean (Euro-Med)—English, n.d.). Another mitigation strategy is to engage in the mitigation of coastal erosion and the restoration of beach areas (Climate Adapt, n.d.). By using these approaches the groundwork is set for future regional policies to improve land sustainability and water scarcity.

Case 4: Eastern Europe

Introduction—According to the IPCC Fifth Assessment Report, Eastern Europe has experienced high levels of permafrost melting, a rise in precipitations, and increased droughts and wildfires (Pietrapertosa et al., 2018). This region is also vulnerable to climate change, as it is experiencing severe impacts on agriculture, water and forest. Climate migration is the aftermath of these agricultural impacts, which has resulted in roughly 142,000 deaths over the last 40 years. Climate refugees are forced to cross the borders, which creates problems for host countries as they are not equipped to take them (Noonan & Rusu, 2022).

Challenges—Moldova is a vulnerable country to climate change whose economy is dependent on agriculture. Moldova experienced nine droughts between 1990 and 2007. In 2007 roughly 80% of the country was covered in water, which impacted the economy. Data from the hydrometeorological services of Ukraine, Belarus and Moldova showed the hottest summer in the last 100–120 years was in 2010 (Nikolayeva, 2012). Climate change is also affecting health in the urban areas of Eastern Europe, which houses two-thirds of the population. Health in impacted urban areas includes heat waves and water scarcity (Pietrapertosa et al., 2018), forcing people to migrate from rural areas.

Mitigation and Adaptation—Out of 885 cities in Eastern Europe, only 147 considered both mitigation and adaptation strategies (Grafakos et al., 2020). According to the Southeast European (SEE) OrientGate Project study, 7 of 11 countries in Eastern Europe have implemented National Adaptation Strategies (NAS) and National Action Plans (Grafakos et al., 2020). Moldova has implemented an intervention that encourages the use of biomass produced from agricultural waste and solar energy, which has reduced carbon emissions to 70,000 tons per year (UNDP, 2023). Using the international Covenant of Mayors for Climate and Energy, Eastern Europe aims to reduce carbon emissions by at least 40% by 2030. This initiative also focuses on building networks between cities in order to raise awareness. Forest and agriculture adaptation strategies introduced in Eastern Europe include afforestation, sustainable forest management and forest fire suppression. In addressing the impact of water and droughts, these initiatives seek to develop flood risk maps, management plans, improvement of areas prone to flooding and improving water quality and monitoring. The initiative considers building models from Linz, Australia which are made with green roofs improving insulation from heat, cold, noise and pollution (Pietrapertosa et al., 2018). To further reduce greenhouse emissions, urban policymakers must connect to climate change policies using a bottom-up approach (Grafakos et al., 2020). Mitigation strategies must prioritize decarbonizing the energy sector, decreasing non-carbon greenhouse gas emissions from agriculture and increasing carbon sinks such as afforestation (Pietrapertosa et al., 2018). All of these initiatives can prevent forced migration from rural areas.

Case 5: Horn of Africa

Introduction—The effects of climate change are felt most strongly in countries in the Horn of Africa, such as Djibouti, Eritrea, Ethiopia and Somalia (Bedasa & Bedemo, 2023). The economy in countries mostly depends on agriculture, which has been severely impacted by inadequate seasonal rainfall, resulting in food insecurity (Seife, 2021). The rise in temperature as a result of climate change, created famine and drought in the region (Azadi et al., 2022), forcing many to leave their homes.

Challenges—Climate change has made Djibouti's environment harsh and its natural resources scarce. Djibouti has been dealing with higher temperatures and dry land increasing food insecurity (Ozer & Mahamoud, 2013). Since Djibouti had to import all of its food, poverty and migration increased. The rising sea level in Djibouti, increased erosion and lost habitat for fish, birds and plants. The landscape in Djibouti is unsuitable for agriculture with 89% of the country being very dry without enough rain for plants to grow causing continuous drought and famine, increasing migration. Migration is showing a rise in Djibouti as a result of greater unemployment, a shortage of clean water and poor hygiene (Ozer & Mahamoud, 2013).

Although Eritrea has access to the red sea coast, climate change and inadequate infrastructure slow agriculture growth. Political instability and war with neighbouring Ethiopia have been the main sources of migration in Eritrea. In recent years severe drought and famine because of climate change have led to large migration. Eritrea has also been impacted by rising temperatures and poor seasonal rainfall exacerbating migration outflow (Ghebrezgabher et al., 2016).

Ethiopia's economy depends on agriculture. The increase in temperature, decrease in precipitation and unprecedented seasonal rainfall increase food insecurity (Mohammed, 2020). Ethiopia has abundant water and natural resources which have been impacted by climate change. The country has been dealing with drought, famine, poverty and unemployment, increasing migration (Asefawu & Nedessa, 2022).

Somalia's main source of water is neighbouring Ethiopia. Somalia's agriculture has been in danger because of the construction of dams in Ethiopia for hydropower, which increases food insecurity leading to migration. The agriculture industry has been affected by higher temperatures and unseasonal rainfall increasing food insecurity (Conway, 2011; Jimale et al., 2023).

Mitigation and Adaptation—Improving food insecurity in the Horn of Africa requires extensive mitigation and adaptation strategies (Lobell et al., 2008). Mitigation and adaptation strategies can be achieved through an integrated approach. Farmers, non-profit organizations working on climate change mitigation, government entities and the international community are required to work in collaboration to eradicate drought and famine. Countries in the Horn of Africa are working to mitigate climate change effects by integrating water and land management, diversifying livelihoods, assessing agriculture with technology and advanced infrastructure and integrating traditional and scientific approaches (UNFCCC, 2022). Countries in this region are engaging prone-migration individuals in advancing the agriculture industry and integrating technology to enhance farming. This government strategy opens an opportunity for unemployed individuals to become entrepreneurs and eradicate poverty and migration.

Discussion of Case Studies and Global Solutions

This section includes a summative discussion of the cases and an analysis of global climate change solutions.

Summative Discussion of Cases

The case studies presented suggest that the examined regions face similar and different climate impacts; the regions have diverse migration patterns, and adaptations and mitigation strategies. Yet, regions face extreme weather events and environmental disasters compounded with political instability and substandard socio-economic conditions, forcing many people to leave their homes. To cope with the uncertainty of climate change, people already living with socio-economic or other vulnerabilities choose to migrate within their borders or to nearby countries. In quest for survival, displaced populations must seek refuge away from their familiar communities, regions and sometimes neighbouring countries.

As analysed through case research, climate change exacerbates environmental disasters, leading to forced migration. Massive migration contributes to further political, economic, or social crises in the countries of origin, transit and destination. These crises are compounded with internal conflicts and climate change. These multi-directional dynamics make the relationship between climate change and human migration difficult to understand and analyse in order to formulate, implement and evaluate policy options, strategies and interventions. The case studies have also examined mitigation and adaptation strategies in five world regions. These strategies paired with adequate migration policies are essential to address forced displacement within and outside borders.

In countries impacted by climate change as well as endemic conflicts and structural oppression, assistance and humanitarian aid programmes are critical. Particularly those aimed at mitigating the impact of natural disasters and promoting community- and national-level adaptation strategies to endure climate change. Case study research evidence suggests that armed conflicts and internal displacements are interconnected to famine, floods and drought, as well as other natural disasters, which in turn increase migration and aggravate the world's refugee crises. In complex contexts, national, regional and global, efforts must be oriented towards mitigating climate disasters and promoting adaptation in the face of climate change while putting people at the centre of policy actions. Good governance is crucial to provide long-term stability and the development of inclusive and sustainable solutions for addressing climate change. As discussed next, these outcomes can be best achieved through strategies aimed at climate resilience, developed and implemented through climate governance.

Climate Change Solutions

As examined through the case studies from five regions of the world, national and regional solutions can only go thus far. Global solutions are necessary in order to address the global challenges faced with climate change and its nexus to climate migration. Climate reliance and climate governance are essential for more effective development of mitigation and adaptation strategies.

Climate resilience can be understood as the ability (of humans in interaction with the natural environment) to rebound from (climate) adversity. Climate resilience as a policy action is conceptualized on the understanding of "the interdependence of climate, ecosystems and biodiversity, and human societies" (Intergovernmental Panel on Climate Change [IPPC], 2022, Summary for Policymakers, p. 5). The IPCC (2022) proposes a shift from adaptive strategies that have many limitations or produce maladaptation to climate resilience development for natural and human systems; to this end, human health and wellbeing, as well as equity and justice, are placed at the centre of mitigation efforts, and ecosystems health, as well as planetary health become attainable goals. The climate change phenomena and contribution to climate change disasters and human migration must adopt approaches that protect and safeguard climate-displaced persons. More inclusive goals can be achieved by acknowledging the nexus between climate change and climate migration by centring climate resilience in all mitigation and adaptation strategies.

Delays in adopting and implementing gold standard climate change policies at the local level may affect the effective protection and humanitarian assistance to internal climate refugees. In the presence of governance failures at the local level, climate governance helps to develop capacity by creating incentives for country-level institutions to develop climate policy (The World Bank, n.d.). Climate governance must involve central governments, local and national governments, as well as international bodies, the private sector and civil society, which includes environmental and civic organizations, youth and women's groups, non-governmental organizations (NGOs) and private voluntary organizations (PVOs), communities affected by climate change and many other grassroots and non-profit organizations. Climate governance, such as government transparency and citizens holding governments accountable, could enable the effective climate action (The World Bank, n.d.), which may include risk reduction and collaborative assistance programmes.

Many international and national organizations promote climate governance because the climate crisis has become a global humanitarian crisis and transnational collaborative climate action to address these crises is urgently needed (UN, 2021). Dialogue surrounding these intersecting crises within the context of climate governance can lead to transformation and the promotion of more viable eco-human mitigation and adaptation strategies aimed at climate resilience. Multi-level climate governance is about engaging a wide range of social actors in discussion and decision-making processes that are formal or informal, as well as flexible and adaptive. Climate change solutions must ensure coherence, collaboration and integration in the identification of structures and mechanisms as well as goals, roles, relationships and responsibilities, among others (UNDP, n.d.). The involvement of worldwide social actors, not just governments, in international climate change fora is fundamental for the development of inclusive and enduring climate change solutions.

Study Implications

The case studies identified challenges and opportunities in areas of unmet needs of refugees and immigrants displaced by climate change. Professionals serving this population, such as social workers, play a critical role in promoting the wellbeing and reducing disparities of the most vulnerable population. The interconnectedness of environmental and human sciences provides an opportunity to further address social and environmental justice for immigrants and refugees experiencing climate change. Interagency collaboration offers an opportunity to mitigate solutions for climate change while advocating and serving environmental refugees (Dorn, 2019).

Social workers, as well as other professionals serving refugees and immigrants, can play an important role in safeguarding vulnerable climate refugees. Social workers and other professionals may serve refugees at the local, national and global level, hence play a critical role in identifying and meeting the needs of climate refugees. The educational curriculum of these professionals must better prepare them for working with refugees and immigrants affected by climate change. Professionals serving this population must be trained in implementing culture sensitive interventions considering the cultural beliefs of climate refugees (George, 2012) and attending their unique needs.

Conclusion

As evident in the case studies examined here, national policies often undermine alternative solutions for climate-affected people; they may also impair peoples' ability to rebuild their cultures and communities in territories of their choice. We have examined some promising mitigation and adaptation strategies that have potential to bring about change in local communities and discourage further climate migration. We have also discussed innovative approaches grounded on international partnerships and community-based initiatives aimed at addressing both climate change and climate migration. We found that global, long-lasting solutions require centring mitigation and adaptation strategies on achieving human ecosystems as well as planetary health (climate resilience), and the involvement of a wide range of social actors (climate governance). Understanding the impact of climate change on their wellbeing is an important precursor for the development of effective policies and practices. Decision-makers and professionals around the world must acknowledge the link between climate change and climate migration while adopting and implementing policies and practices protecting and empowering climate migrants.

References

Alam, J. (2022). For climate migrants, Bangladesh offers promising alternatives. *The Diplomat*. https://thediplomat.com/2022/03/for-climate-migrants-bangladesh-offers-promising-alternatives/

Apap, J. (2021). The concept of 'climate refugee': Towards a possible definition. *Think Tank, European Parliament*. https://www.europarl.europa.eu/thinktank/en/document/EPRS_BRI(2021)698753

Asefawu, G. S., & Nedessa, K. G. (2022). Nature, patterns, and determinants of seasonal outmigration in the North-Eastern part of Ethiopia. *Norsk Geografisk Tidsskrift - Norwegian Journal of Geography, 76*(3), 177–191. https://doi.org/10.1080/00291951.2022.2076610

Azadi, H., Burkart, S., Movahhed Moghaddam, S., Mahmoudi, H., Janečková, K., Sklenička, P., Ho, P., Teklemariam, D., & Nadiri, H. (2022). Famine in the Horn of Africa: Understanding institutional arrangements in land tenure systems. *Food Reviews International, 38*(sup1), 829–845. https://doi.org/10.1080/87559129.2021.1888974

Barbier, E. B., & Hochard, J. P. (2018). The impacts of climate change on the poor in disadvantaged regions. *Review of Environmental Economics and Policy, 12*(1), 26–47. https://doi.org/10.1093/reep/rex023

Bedasa, Y., & Bedemo, A. (2023). The effect of climate change on food insecurity in the Horn of Africa. *GeoJournal, 88*(2), 1829–1839. https://doi.org/10.1007/s10708-022-10733-1

Climate Adapt. (n.d.) *Mediterranean (Euro-Med)*. https://climate-adapt.eea.europa.eu/en/countries-regions/transnational-regions/mediterranean#:~:text=The%20main%20goal%20of%20the,the%20well%2Dbeing%20of%20its

Coda, J. C. (2021). Promoting climate change action in Latin America and the Caribbean. *World Bank*. https://www.worldbank.org/en/results/2021/04/14/promoting-climate-change-action-in-latin-america-and-the-caribbean

Conway, D. (2000). Some aspects of climate variability in the north east Ethiopian highlands—Wollo and Tigray. *SINET: Ethiopian Journal of Science, 23*(2), 139–161. https://doi.org/10.4314/sinet.v23i2.18163.

Conway, D. (2011). Adapting climate research for development in Africa. *Wiley Interdisciplinary Reviews: Climate Change, 2*(3), 428–450. https://doi.org/10.1002/wcc.115

Cherif, S., Doblas-Miranda, E., Lionello P, Borrego, C., Giorgi, F., Iglesias, A., Jebari, S., Mahmoudi, E., Moriondo, M., Pringault, O., Rilov, G., Somot, S., Tsikliras, A., Vila, M., Zittis, G. (2020). Drivers of change. In W. Cramer, J. Guiot, & K. Marini, K. (Eds.), *Climate and environmental change in the Mediterranean basin—current situation and risks for the future. First mediterranean assessment report* (pp. 59–180). Union for the Mediterranean, plan Bleu, UNEP/MAP, Marseille, France, https://doi.org/10.5281/zenodo.7100601.

Crate, S. A. & Nuttall, M. (2016). *Anthropology and climate change: From Actions to Transformations (2nd Ed).* Routledge. eBook ISBN9781315530338

Crowe, S., Cresswell, K., Robertson, A., et al. (2011). The case study approach. *BMC Medical Research Methodology, 11*, 100. https://doi.org/10.1186/1471-2288-11-100

Dorn, C. (2019). Climate change and health: A call to social workers. *Summer practice perspectives*. https://www.socialworkers.org/

Environmental protection agency. (2022). Climate change and the health of socially vulnerable people. *Environmental protection agency (EPA)*. https://www.epa.gov/climateimpacts/climate-change-and-health-socially-vulnerable-people

Environmental protection agency. (2023). Climate change indicators: Wildfires. *Environmental protection agency (EPA)*. https://www.epa.gov/climate-indicators/climate-change-indicators-wildfires

European environment agency. (2022). What is the difference between adaptation and mitigation? Retrieved July 5, 2023 from https://www.eea.europa.eu/help/faq/what-is-the-difference-between

George, M. (2012). Migration traumatic experiences and refugee distress: Implications for social work practice. *Clinical Social Work Journal, 40*(4), 429–437. https://doi.org/10.1007/s10615-012-0397-y

Ghebrezgabher, M. G., Yang, T., & Yang, X. (2016). Long-term trend of climate change and drought assessment in the horn of Africa. *Advances in Meteorology, 2016*, 1–12. https://doi.org/10.1155/2016/8057641

Gibbert, M., & Ruigrok, W. (2010). The '"what"' and '"how"' of case study rigor: Three strategies based on published work. *Organizational Research Methods, 13*(4), 710–737. https://doi.org/10.1177/1094428109351319

Giri, S., Prabhakar, A., Malla, R. B., Oli, S., Poudel, S., & Khanal, A. (2023). Climate change mitigation and adaptation in Nepal and South Asia: Challenges, progress, and recommendations. *Journal of Sustainability and Environmental Management, 2*(2), 133–140. https://doi.org/10.3126/josem.v2i2.55206

Grand challenges for social work. (2021). About. Retrieved June 20, 2023, from https://grandchallengesforsocialwork.org/create-social-responses-to-a-changing-environment/

Grafakos, S., Viero, G., Reckien, D., Trigg, K., Viguie, V., Sudmant, A., Graves, C., Foley, A., Heidrich, O., Mirailles, J. M., Carter, J., Chang, L. H., Nador, C., Liseri, M., Chelleri, L., Orru, H., Orru, K., Aelenei, R., Bilska, A., & Pfeiffer, B. (2020). Integration of mitigation and adaptation in urban climate change action plans in Europe: A systematic assessment. *Renewable and Sustainable Energy Reviews, 121*, 109623. https://doi.org/10.1016/j.rser.2019.109623

Grandi, F. (2021). *Climate change is an emergency for everyone, everywhere.* https://www.unhcr.org/news/stories/climate-change-emergency-everyone-everywhere

Intergovernmental panel on climate change. IPCC. (2022). *Climate change 2022: Impacts, adaptation and vulnerability.* https://ipcc.ch/report/ar6/wg2/

International Monetary Fund. (2020). World Economic Outlook, October 2020: a Long and Difficult Ascent. *IMF.* https://www.imf.org/en/Publications/WEO/Issues/2020/09/30/world-economic-outlook-october-2020

Institute of economics and peace. (2020). Conflict management, peace economics and development. *Ethnic Conflict, Civil War and Cost of Conflict*, ii. https://doi.org/10.1108/s1572-8323(2011)0000017019

Jimale, A. D., Abdullahi, M. O., Ahmed, Y. A., Nageeye, A. Y., Abdullahi, B. S., & Jama, A. A. (2023). Mitigating the impact of floods: An IoT-driven monitoring and alert system for Somalia's rivers. *International Journal of Electrical and Electronics Engineering*, *10*(6), 120–125. https://doi.org/10.14445/23488379/IJEEE-V10I6P113

Krishnamurthy, P. K., Lewis, K., Kent, C., & Aggarwal, P. (2015). Climate impacts on food security and livelihoods in Asia: A review of existing knowledge. https://documents.wfp.org/stellent/groups/public/documents/newsroom/wfp281745.pdf?_ga=2.52639400.1657381638.1678465716-1183756098.1678465716

Lobell, D. B., Burke, M. B., Tebaldi, C., Mastrandrea, M. D., Falcon, W. P., & Naylor, R. L. (2008). Prioritizing climate change adaptation needs for food security in 2030. *Science, 319*(5863), 607–610. https://doi.org/10.1126/science.1152339

Mall, R. K., Srivastava, R. K., Banerjee, T., Mishra, O. P., Bhatt, D., & Sonkar, G. (2018). Disaster risk reduction including climate change adaptation over South Asia: Challenges and ways forward. *International Journal of Disaster Risk Science, 10*(1), 14–27. https://doi.org/10.1007/s13753-018-0210-9

Marchand, M. H. (2021). The *caravanas de migrantes* making their way north: Problematising the biopolitics of mobilities in Mexico. *Third World Quarterly, 42*(1), 141–161. https://doi.org/10.1080/01436597.2020.1824579

McMahon, S., & Sigona, N. (2021). Death and migration: Migrant journeys and the governance of migration during Europe's "Migration crisis." *International Migration Review, 55*(2), 605–628. https://doi.org/10.1177/0197918320958615

Mohammed, B. U. (2020). Review on impact of climate change on Ethiopia agriculture: Current and future prospect. *Journal of Bio Innovation, 9*(5), 750–763. https://doi.org/10.46344/JBINO.2020.v09i05.10

Monico, C. & Pitts, W. J. (2021). Opportunities and challenges affecting migration in the Northern Triangle. *Georgetown Journal of International Affairs*. https://gjia.georgetown.edu/2021/08/12/opportunities-and-challenges-affecting-migration-in-the-northern-triangle/

National Aeronautics and Space Administration. (NASA). (2023). *Overview: Weather, Global Warming and climate change*. NASA. https://climate.nasa.gov/what-is-climate-change/

Naser, M. M. (2012). Climate change, environmental degradation, and migration: A complex nexus. *William & Mary environmental law & policy review*, 36 (713–755), https://scholarship.law.wm.edu/wmelpr/vol36/iss3/4

Nikolayeva, L. (2012). *Climate change in Eastern Europe Belarus, Moldova, Ukraine*. https://www.droughtmanagement.info/literature/ZOI_climate_change_eastern_europe_2012.pdf

Noonan, E., & Rusu, A. (2022). At a glance: Thinking about tomorrow. The future of climate migration. *European Parlament*. https://www.europarl.europa.eu/RegData/etudes/ATAG/2022/729334/EPRS_ATA(2022)729334_EN.pdf

Ozer, P., & Mahamoud, A. (2013). Recent extreme precipitation and temperature changes in Djibouti City (1966–2011). *Journal of Climatology, 2013*, 1–8. https://doi.org/10.1155/2013/928501

Pietrapertosa, F., Khokhlov, V., Salvia, M., & Cosmi, C. (2018). Climate change adaptation policies and plans: A survey in 11 South East European countries. *Renewable and Sustainable Energy Reviews, 81*(2), 3041–3050. https://doi.org/10.1016/j.rser.2017.06.116

Riguad, K., Kanta, de Sherbin, A., Jones, B., Bergmann, J., Clemente, V., Ober, K., Schewe, J., Adamo, S., McCusker, B., Heuser, S., & Midgley, A. (2018). The World Bank. Groundswell: Preparing for internal climate migration. *The World Bank*. https://openknowledge.worldbank.org/bitstream/handle/10986/29461/WBG_ClimateChange_Final.pdf

Seife, T. K. (2021). The impact of climate change on agriculture and food security in the greater Horn of Africa. *Politikon, 48*(1), 98–114. https://doi.org/10.1080/02589346.2020.1861509

Saidi, K., & Omri, A. (2020). The impact of renewable energy on carbon emissions and economic growth in 15 major renewable energy-consuming countries. *Environmental Research, 186*, 109567. https://doi.org/10.1016/j.envres.2020.109567

Shultz, J. M., Rechkemmer, A., Rai, A., & McManus, K. T. (2018). Public health and mental health implications of environmentally induced forced migration. *Disaster Medicine and Public Health Preparedness, 13*(02), 116–122. https://doi.org/10.1017/dmp.2018.27

Smith, A. B. (2023). 2022 U.S. billion-dollar weather and climate disasters in historical context. *NOAA Climate.gov*. https://www.climate.gov/news-features/blogs/2022-us-billion-dollar-weather-and-climate-disasters-historical-context

Suarez, I. (2020). 5 strategies that achieve climate mitigation and adaptation simultaneously. *World Resources Institute*. Retrieved January 5, 2023, from https://www.wri.org/insights/5-strategies-achieve-climate-mitigation-and-adaptation-simultaneously

Thant, P. S., Espino, A., Soria, G., Myae, C., Rodriguez, E., Barbon, W. J., & Gonsalves, J. (2022). Myanmar local food systems in a changing climate: Insights from multiple stakeholders. *Environmental and Sustainability Indicators, 14*, 100170. https://doi.org/10.1016/j.indic.2022.100170

The World Bank. (n. d.). Climate change governance—Overview. *The World Bank*. https://www.worldbank.org/en/programs/climate-change-governance/overview

United Nations (UN). (2021, April 29). The climate crisis is a humanitarian crisis. *United Nation*s. Retrieved January 4, 2023, from https://www.un.org/en/climatechange/the-climate-crisis-is-a-humanitarian-crisis

United Nations (UN). (2022, November 20). COP27 reaches breakthrough agreement on new "loss and damage" fund for vulnerable countries. Retrieved on November, 12, 2024 from https://unfccc.int/news/cop27-reaches-breakthrough-agreement-on-new-loss-and-damage-fund-for-vulnerable-countries

United Nations Climate Action. (2023). What is climate change?. *United Nations*. https://www.un.org/en/climatechange/what-is-climate-change

United Nations Development Programme. (UNDP). (2023). *Climate change, environment & energy* https://www.undp.org/moldova/climate-change-environment-energy

United Nations Development Programme. (UNDP). (n.d.). Toolkit for young climate activists in the Middle East and North Africa Region/Arab States Region. Volume III: What is climate governance? https://www.undp.org/sites/g/files/zskgke326/files/2022-09/Volume%20III%20What%20is%20climate%20governance.pdf

United Nations Environment Programme. (1985). Environmental refugees. ISBN: 92-807-1103-2. Retrieved on June 20, 2023 from the UN Digital Library. https://digitallibrary.un.org/record/121267?ln=en

United Nations. (2023). Human rights-based approach to migration. *Office of the United Nations High Commissioner for Human Rights (OHCHR)*. https://www.ohchr.org/sites/default/files/documents/publications/Towards-Human-Rights-Based-Approach-Migration-Training-Guide.pdf

United Nations High Commissioner for Refugees. (UNHCR). (2023). Refugee statistics. *UNHCR*. Retrieved June 20, 2023, from https://www.unhcr.org/refugee-statistics/

United Nations framework convention on climate change. (UNFCCC). (2022). The Paris agreement. *UNFCCC*. https://unfccc.int/process-and-meetings/the-paris-agreement

White House. (2021). Report on the impact of climate change on migration. *Reliefweb*. https://reliefweb.int/report/world/report-impact-climate-change-migration-october-2021

PART IV

Vulnerable Populations and Climate Modelling (AI)

CHAPTER 11

Why Climate Change Matters for Vulnerable Populations in South Africa: A Social Work Perspective

Tamsin Leigh Nel and Priscalia Khosa

Introduction

Climate change is not seen as an isolated crisis, but as part and consequence of interrelated issues, such as industrialisation and urbanisation. For this chapter, climate change is understood as "the trend of changes in the earth's general weather conditions as a result of an average rise in the temperature of the earth's surface often referred to as global warming", (Department of Environmental Affairs, 2012). According to Pierrehumbert (2019), the warming of the earth is due to an excess emission of greenhouse gases (GHG), which are largely contributed by what is known as anthropogenic factors, namely human activity, such as the fossil-fuelled

T. L. Nel (✉)
Stellenbosch University, Stellenbosch, South Africa
e-mail: tamsinnel01@gmail.com

P. Khosa
University of Johannesburg, Johannesburg, South Africa
e-mail: priscaliak@uj.ac.za

D. D. Joseph et al. (eds.), *Climate Change*,
https://doi.org/10.1007/978-3-031-83037-2_11

economy and industrial processes. Although the Earth's atmosphere can absorb GHG emissions due to geological processes, the concern is the rate at which these gases are emitted into the atmosphere that erodes the earth's natural defences and geological processes (Pierrehumbert, 2019).

The entirety of the effects of climate change are still speculative and not fully comprehended, but evidence has demonstrated that climate change is challenging various domains of society. In terms of the physical environment, catastrophic environmental events such as droughts, extreme temperatures and floods are becoming more unpredictable (Boddy & Hetherington, 2013; Kompas et al., 2018). In terms of societal experiences, social challenges may also occur because of the effects of climate change, such as food insecurity, forced migrations and health concerns, all of which influence the livelihoods of communities and their physical as well as social functioning (Dominelli, 2011). With its focus on person-in-environment approach, the social work profession values the worth and dignity of human beings, and the profession has increasingly been called upon to respond to the effects of climate change. Failure to respond to the call as a profession will be a missed opportunity to play a vital role in assisting communities that are already marginalised, which may be exacerbated by the effects of climate change (Boddy & Hetherington, 2013). To effectively respond to the welfare of communities, social workers need to understand climate change and its effects, but this relationship is often unrecognised, requiring a thorough investigation to inform social work as a profession.

Climate change is not synonymous with one area or country, but it is a global crisis, and although many scientists agree it is real through evidence-based reports, there are still climate change sceptics (Boddy & Hetherington, 2013). The South African government has recognised that climate change does in fact exist and has postulated various effects it may have upon the country (Department of Environment, Forestry & Fisheries, 2019). Many strategic plans, laws and approaches have been created in South Africa such as the National Climate Change Response Plan White Paper (NCCRP) of 2012 and the Disaster Management Amendment Act, 2015 (Act No. 16 of 2015). According to the National Climate Change Response Plan White Paper (NCCRP), the South African government is committed to climate change responses with an overall approach to be needs driven, developmental, empowering, transformational, participatory, integrated and evidence-based (Department of Environmental Affairs, 2012). The two main objectives of the NCCRP are: (1) to better South Africa's adaptation to climate change through the country's economic, environmental and social resilience; and (2) to mitigate climate change through contributing to the global effort to level out GHG concentrations (Garland, 2014). Another significant stride towards directly responding to climate change in South Africa is the Disaster Management Amendment Act, 2015 (Act No. 16 of 2015).

The Disaster Management Amendment Act, No 16 of 2015, assigns responsibility to local, provincial and national spheres of government to invest in climate change adaptation interventions and disaster risk mitigation. In essence, an environmental justice position can be followed, which emphasises human rights in relation to the environmental degradation to reduce environmental harm and ensure equitable access to natural resources (Morgaine & Capous-Desyllas, 2020). By making use of the environmental justice position, together with the use of existing legislation like the Disaster Management Amendment Act, social workers in South Africa can advocate for human rights in respect to environmental concerns and develop best practices to address societal challenges associated with climate change. However, for social workers to respond from an informed practice in terms of climate change, the effects upon vulnerable groups need to be better understood.

This chapter provides a critical analysis of the impacts of climate change on vulnerable populations, analysing the issue from both a South African context and a social work perspective. More specifically, the vulnerable groups highlighted are those living in poverty and women. A remit to action from a social work perspective is suggested as these vulnerable groups are the very clientele for social workers. Based on a critical review, the chapter will begin by conceptualising climate change from a global and South African context. The discussion provides an in-depth examination of the basis upon which social work should respond to climate change. This will be followed by the effects of climate change in society by focusing particularly on vulnerable groups. To conclude, implications and recommendations for social work education and practice are provided.

Climate Change: A Global and South African Context

Climate change is a global problem that needs to be treated through cooperative efforts from all countries. Consequently, to limit the temperature increase to 2 °C or less from a global standpoint, government accords and conferences, such as the Intergovernmental Panel on Climate Change (IPCC), are significant in bringing countries together to create collective responses. In addition, international agreements exist to ensure that at a global level, countries are bound to their commitment to reduce their emissions and combat climate change. One such agreement is the Kyoto Protocol, initially adopted by 35 countries in 1997. However, it has been highly criticised for not deterring non-compliance, and studies on the efficiency of the Kyoto Protocol indicate that there were no provable effects on the decline of carbon dioxide emissions from the signatory countries, South Africa being one of them, but a continuing global increase instead (Almer & Winkel, 2017). Moreover, the Kyoto Protocol recognises developed countries as the largest contributors for GHG emissions and only legally binds the developed countries, creating a limitation. To address the shortcomings of the Kyoto Protocol, the Paris Agreement in 2015 offered developing countries an opportunity to work alongside developed countries in the fight against global warming (McCloy, 2020). These agreements are a pledge by signatory countries to collectively slow down global warming in their respective nations, yet records show that the GHG emissions rose an estimated 2.7% since 2018 globally (Pierrehumbert, 2019).

Given the rise in GHG emissions, all continents have already experienced changes in natural systems through extreme weather events and an increase in sea levels, with the added threat to agricultural production in many parts of the world (Boddy & Hetherington, 2013; Kompas et al., 2018). Effects of climate change will be compounded by water shortages with 80% of the world's population likely to face extreme water insecurity. Human health and wellbeing are also at risk due to global warming, in which evidence has shown a direct link between global temperature increments with increased risk for human health manifesting in diseases such as malaria, dengue and diarrhoea. Furthermore, it was estimated that global warming already causes 300,000 deaths annually (Kompas et al., 2018).

Based on geography and economics, some regions will experience opportunities from climate change whilst others will undergo increased vulnerabilities (Boddy & Hetherington, 2013). The impact of global warming varies by time, region and economic sectors, but trends show that the devastation in relatively poor African and Asian nations worsen over time (Kompas et al., 2018; Niang et al., 2014). These two continents are of particular importance given that they are the two-largest continents in geographical size and population. In Asia, observable and projected impacts of climate change will affect water availability and quality, both of which are anticipated to decline (Hijioka et al., 2014).

The African continent has been deemed as the most vulnerable continent due to its predicted high exposure to the effects of climate change and existing low adaptive capacity (Chersich et al., 2018). Although Africa's vulnerability source is mainly socio-economic in nature, there are observable and projected trends that depict temperature increases and variability as a major concern. Trends show that not only has Africa experienced a warming beyond the natural variability range, but temperatures are projected to continue to increase faster than that of the global average increase (Niang, 2014). For the Southern African region, this is no different as land surface temperatures are likely to increase in all-weather seasons. These trends are of particular concern given that water availability will be negatively impacted, which only amplifies the existing water-stressed concern for most countries in the continent. In addition, temperature increases will reduce agricultural productivity and thus lead to food insecurity (Niang, 2014). Subsequently, the continent's existing economic development constraints will be exacerbated, together with other human security issues such as health care and food insecurity (Niang, 2014; Chersich et al, 2018).

South Africa is vulnerable to climate change and has low capacity to cope with the impacts thereof due to its environmental and socio-economic context. It is also recognised as a significant contributor to climate change through elevated levels of GHG emissions (Department of Environmental Affairs, 2012). These factors compel South African social workers to be more aware of the ways in which environmental factors can affect the wellbeing of the vulnerable communities they serve. For example, the City of Cape Town in the Western Cape Province of South Africa experienced a severe drought between 2015 and 2018. The lack of rainfall meant that the city imposed strict water restrictions due to the probability of a Day Zero of no water. Though this was narrowly avoided, the irony that a coastal city could have run out of water was improbable in the past (Swain, 2019). Juxtapose extreme rainfall in KwaZulu-Natal province in April 2022 led to catastrophic landslides and floods where large parts of the province were left without electricity, water, damaged infrastructure and thousands of livelihoods negatively affected (Pinto et al., 2022). These examples demonstrated the accumulative effects of climatic events on the livelihoods of communities and sustainable development issues such as health care, employment opportunities, food security and housing (Pinto et al., 2022). This is coupled by the South African economy, which is currently unstable, as the country was estimated to have the highest Gini Coefficients (global measure of inequality) in the world which perpetuates inequality and exclusion of the most vulnerable groups (Department of Environment, Forestry & Fisheries, 2019; Chersich et al., 2018). The inequality gap between the poor and rich continues to widen and it is exacerbated by multiple factors. First, a significant number of South Africans are unemployed. According to the latest statistics, the unemployment rate in South Africa was estimated at 32.6% in June 2023. Of the 16.4 million employed, only 11.3 million works in the formal sector, where incomes are typically higher. These numbers demonstrate that the economy is unable to generate enough employment opportunities. The second factor is that many of those who are employed earn very low wages. Around 3 million people work in the informal economy, where incomes are generally low. Additionally, approximately 900,000 people are employed in agriculture, and about 1 million are domestic workers, both sectors characterised by low incomes (Valodia, 2023).

Over time, social work has increasingly emphasised its role in empowering marginalised individuals to voice their concerns, which might otherwise go unnoticed (Masoga & Shokane, 2019). There exist foundational principles through which social work can integrate environmental considerations into its practice, with the ecological perspective standing out as particularly relevant. This perspective acknowledges the intricate web of interactions and dependencies amongst organisms and their surroundings, highlighting humans as integral components within a broader ecosystem. Furthermore, it embraces the notion of 'person-in-environment', recognising that human behaviour is shaped not just by internal factors but also by external influences from both physical and social surroundings. Despite drawing insights from ecology to understand how individuals are shaped by their environments, the social work field traditionally focused more on cultural and social contexts and less attention to the natural environment's impact on human wellbeing (Coates & Gray, 2012; Teater, 2014). It's therefore crucial for the profession to acknowledge the significance of the natural environment by empowering communities to preserve and enhance it for future generations (Masoga & Shokane, 2019). This underscores the importance of recognising the connection between social work and climate change.

The Relationship Between Social Work and Climate Change

To understand the link between social work and climate change, it is valuable to stipulate the global social work definition, which was approved by the International Federation of Social Work General Meeting, and the International Association of Schools of Social Work (IASSW) General Assembly in July 2014. It is understood as a:

> Practice-based profession and an academic discipline that promotes social change and development, social cohesion, and the empowerment and liberation of people. Principles of social justice, human rights, collective responsibility, and respect for diversities are central to social work.

Based on this definition, the positive wellbeing of client systems is of utmost importance. Although climatic events may result in large-scale economic outcomes for communities and families, social impacts may also result in devastating costs. Such social effects can include water and food shortages; homelessness and displacement; and increase in mortality rates amongst vulnerable groups such as those with poor living and health conditions (Bowles et al., 2018). Social workers working with vulnerable groups that are disproportionally affected by climate change have a place to intervene by addressing its effects by enhancing people's wellbeing and actualise the social justice goal of the profession (Dominelli, 2012; Masoga & Shokane, 2019). The notion that climatic events can cause socio-economic issues as well as devastating outcomes for communities results in the profession being concerned for the welfare of vulnerable populations.

Within the profession, there is a prevailing interrelationship between people and their environment, through the concept of person-in-environment. Germain and Gitterman's (1980) life model introduced the notion of person-in-environment in which people's lives are influenced by how they live, demonstrating social workers' understanding of people's transactional relationship between their social environments. However, this approach only contextualises socio-cultural factors, disregarding the transactional process involved with the natural environment. Herein lies an opportunity for the person-in-environment approach to be expanded, to include the physical environment and to work in a holistic manner to enhance people's social functioning (Rambaree, 2020; Bowles et al., 2018). Furthermore, as a profession, social work engages across micro-, mezzo- and macro-levels of practice and policy advocacy to create social change because of the inherent work within communities to combat complex social issues and crises (International Federation of Social Work, 2022).

When conceptualising the relationship between climate change and social work, it is also important to focus on the commitments made by several social work bodies to respond to climate change and address environmental concerns in practice. The IFSW calls for social workers to become aware of as well as integrate environmental sustainability into practice. Their call states that social work should recognise the alliance between the natural and social world to enhance environmental responsibility and care for the environment for future generations. The IFSW (2022) also reiterates that environmental issues should gain increased presence in social work education. Given that the IFSW represents social workers within an international context and encourages social change from a human rights and social justice perspective, their standpoint on environmental issues is important for social workers (Bowles et al., 2018). Thus, the IFSW commits social workers to the task of being concerned with climate-related issues through social justice and collective responsibility.

With a history of colonialism and suppression enforced during the apartheid era in South Africa, a new approach to welfare was constructed after 1994 when democracy was instated to dismantle inequality and discrimination. As a result, the democratic government enacted a developmental social welfare (DSW) policy through the national White Paper for Social Welfare in order to guide the provision of social welfare services in the country. Developmental social welfare is an evolving approach in South Africa that promotes social development, social justice and collective responsibility. Key ideals of this South African welfare policy include fairness, equity and participation which foregrounds social work practice in meeting the basic needs of all South Africans (Lombard & Twikirize, 2014). Through DSW, multiple social work methods are encouraged, such as the micro-, meso- and macro-levels of intervention, with services across rehabilitation, protection, prevention and developmental goals. DSW advocate for human rights and people-centred theories to improve not only the welfare of the population but economic development (Patel & Hochefeld, 2013). An integrated service delivery model (ISDM) was developed to ensure the implementation of the DSW approach

(Strydom et al., 2020). With these social welfare policy underpinnings, the value of human rights perspective and climate change policies in South Africa, the basis for social work practice is suitable and fair to respond to climate change effects imposed upon communities.

The embracement of social justice and collective responsibility by social workers is evident through their advocacy role in safeguarding human rights. For instance, the socio-economic rights within the South African Constitution include environmental rights where it guarantees (1) the right to an environment that is not harmful to one's wellbeing and health and (2) the right to have one's environment protected (Lombard & Twikirize, 2014). These environmental rights are collective in nature, which is useful to understand social workers' role at a macro level. Social workers play an advocacy role for protecting these rights in order to enhance people's human dignity and engage with key stakeholders such as the state to create basic conditions for the enhancement of socio-economic rights. Hence, social workers' roles also encompass policy development and research. In advocating for environmental rights and sustainable development practices in South Africa, social workers at a macro-level practice are involved in community development initiatives based on communities' needs and climate change policy discussion to uphold social and environmental justice (Lombard & Twikirize, 2014).

Effects of Climate Change on Vulnerable Populations

Those vulnerable in society are more susceptible to the negative impacts of climate change (Boddy & Hetherington, 2013). The United Nations Development Programme (UNDP) postulated that the manner and extent to which communities will be affected by climate change is due to their gender, poverty, socio-economic status and access to resources (Muchacha, 2019). In fact, climate change intensifies inequality. Research on vulnerability to climate change depicts that whilst everyone will be affected, marginalised populations such as people living in rural areas, impoverished communities, the elderly and women will be disproportionally affected, which are the very vulnerable groups social workers serve (Boddy & Hetherington, 2013; Diffenbaugh & Burke, 2019). Thus, it is necessity for social workers to play a role in addressing the effects of climate change on vulnerable populations.

Effects of Climate Change on People Living in Poverty

Poverty is not identified as synonymous with climate change as it is a disaster on its own, but its presence can only exacerbate the problems that may be encountered due to climate change (Dominelli, 2012). It is recognised that those who are well resourced have access to resources in their communities and in choices on where to live, resulting in their ability to choose to live away from environmental risks, and are able to adapt quicker after a devastating climatic event. Hence, poverty stands out as a primary vulnerability stressor, significantly impacting the ability to adapt to climate change.

Poverty often coincides with a lack of environmental rights, trapping people in deteriorating living conditions (Närhi & Matthies, 2016). Individuals grappling with poverty often face significant barriers to accessing vital social and economic resources such as water, food, education, infrastructure, housing and services. It is crucial to recognise that for those already struggling with depleted resources and limited capacities, the effects of climate change can exacerbate their challenges. This may mean that those in poverty may not be able to afford resources to increase their ability to adapt to climate change or cope after a climatic event; may not receive the material or social assistance to either build resilience or adapt after a climatic disaster (Boddy & Hetherington, 2013). Furthermore, poverty frequently intertwines with a reliance on natural resources through agricultural production, which becomes increasingly vulnerable to the impacts of climate change. The unpredictability and severity of extreme weather events and patterns induced by climate change pose a significant threat to natural resources or agricultural production. For example, crop yields could deplete either through climatic disasters or unfavourable weather conditions triggering food insecurity for populations (Walker, 2012).

Within the context of the South African economic state, together with the idea that developing countries are more susceptible to the effects of climate change, it creates cause for concern for the country. The incidence of poverty according to three subjective poverty measures from the latest statistics indicated that 43.3% of South Africans feel the inability to make ends meet. Many South Africans live below the poverty line, with the recent statistical data found between 2022 and 2023 indicating that South Africa's unemployment rate is at 32.9% (Statistics South Africa,

2023b). In addition, between 2022 and 2023, the inflation rates increased for South Africans, whereby the cost of living increased by 11.3% for the poorest households, more than that of the higher expenditure groups. It was also found that the poorest households spent half of their annual budgets on food and non-alcoholic beverages, compared to an 11% spent by wealthy households (Statistics South Africa, 2023a). This demonstrates that the poor bear the brunt of economic struggles and this is exacerbated when a climate event occurs. Moreover, the country experiences a lack in affordable housing and rapid urban growth and so, there is an upsurge of informal settlements in all cities in South Africa (Chersich et al., 2018). An informal settlement in South Africa is an unplanned settlement on land that has not been proclaimed or surveyed as residential, that consists largely of informal dwellings known as shacks, which is a makeshift structure not erected according to approved architectural plans (Housing Development Agency, 2012). Consequently, most poor South Africans reside in low-cost housing and informal settlements, which are poorly insulated and situated in locations that are at high risk of environmental degradation and limited services (Pinto et al., 2022).

Given the complexity of poverty and structural inequalities pertinent in South Africa, it is the responsibility of social workers to understand how those living in poverty may be disproportionally affected by climate change effects in order to engage key stakeholders and facilitate resilience and adaptational capabilities amongst communities. In practice, when assisting individuals living in poverty who may face the impacts of climate change, social workers play a crucial role. This includes conducting assessments of needs and risks, engaging in disaster management efforts, coordinating resources in the aftermath of environmental disasters and offering counselling to those affected by environmental devastation (Dominelli, 2012). Additionally, it's vital to design intervention strategies that prioritise participatory approaches, placing people and their needs at the centre of development, considering their physical, cultural and social contexts (Department of Environmental Affairs, 2012). Social workers also have a significant role to play in advocating for reduction in carbon emissions and promoting clean energy, but also protect the vulnerable populations from the negative impacts of climate change. This plays into the profession being charged with enhancing human wellbeing (Dominelli, 2011).

The Effects of Climate Change on Women

Gender is complex, socially constructed and shaped by different cultural contexts and social customs. Yet it has been argued as a determining factor for vulnerability and resilience to environmental issues worldwide (Alston, 2013). It is worth noting that this chapter does not standardise the experiences of women concerning climate change but considers all the intersections of various issues surrounding gender, age, political, economic and social status, and how these factors may increase or decrease gender inequalities in South Africa. Gender inequality still exists within a large part of the world with restricted land rights, lack of financial resources and limited political decision-making (International Union for Conservation of Nature, 2015). Many people living in poverty in South Africa are women, who also constitute as 51.6% of South Africa's total population (Alston, 2013; Department of Environment, Forestry & Fisheries, 2019; Statistics South Africa, 2022b). In addition, women are less likely to own land and resources, which means that their capacity to recover after disasters is more restricted, and they are less likely to be involved in decision-making and able to express their needs (Alston, 2013). According to a report released by Statistics South Africa, the profile of a subjectively poor household in South Africa is one usually headed by a black African female under 35 years old, who resides in a rural area located in a rural-based province and has a lower level of education. Within these households it is more likely that all economically active individuals (age 15 years and above) are unemployed. On the income distribution, this household is located in the lower quintiles (Statistics South Africa, 2022b).

The United Nations (2022) reports that over 380 million women and girls endure extreme poverty, with projections indicating a potential rise in this figure. By 2030, if present trends persist, the number of women and girls living in poverty in sub-Saharan Africa alone could surpass the current global total (UN Women and United Nations Department of Economic and Social Affairs, 2022). In South Africa, recent consensus data from 2022 revealed that 20.1% of recorded homeless individuals were women, with 41.6% attributing their homelessness to job loss or income reduction (Statistics South Africa, 2022b). However, the notion that women face disproportionate vulnerability to climate change is rooted in systemic oppression and enduring structural inequalities between genders (UN Women and United Nations Department of

Economic and Social Affairs, 2022). In addition, women's susceptibility is rooted from their reduced access to and control over land and environmental goods, the increased likelihood of living in poverty as well as the exclusion from decision-making processes towards adaption and resilience efforts towards climate change. For example, Madzwamuse (2010) posited that 70% of the livelihoods of Zimbabwean women are dependent upon agriculture and rainfall. Therefore, threats to natural resources caused by climate change affecting weather patterns will directly affect the livelihoods of these women in farming.

In addition, women, more specifically from marginalised and poor communities, were disproportionally affected by disasters recorded between 1970 and 2019 (Statistics South Africa, 2022a). Thus, the structural inequalities that persist within society perpetuates women's vulnerability to the effects of climate change (Mulvale, 2017).

Consequently, an eco-feminist approach incorporates a gender sensitivity lens to the issue of climate change from a social work perspective (Boddy & Hetherington, 2013). Eco-feminism brings environmentalism and feminism together to show that the vulnerability of women and environmental degradations are symptoms of capitalism and patriarchy (Mulvale, 2017). Norton (2012) specifically demonstrates that a feminist approach of empowerment and empathy equips social workers to advance an environmental justice perspective when addressing climate change in order to challenge gender disparities and existing unbalanced power relations. This is especially true for South Africa given the existing gender inequality and the high prevalence of femicide (Department of Environment, Forestry & Fisheries, 2019).

In practical terms, social workers can advocate and conduct needs assessments in communities, with the idea in mind that women may be more susceptible to the effects of climate change. In addition, making use of risk reduction strategies to anticipate, reduce and manage the impacts of climate change would assist any vulnerable group (Dominelli, 2012). As a profession dedicated to aiding others, social work must reinforce its pledge to improve the welfare and social integration of communities, recognising that certain groups, particularly the impoverished and women, may face heightened vulnerability to climate change impacts. Nevertheless, there remains ambiguity surrounding a defined environmental agenda within social work, posing difficulties in establishing a conceptual framework for addressing climate change effects (Kemp, 2011).

Implications for Social Work Education and Practice

The implication of understanding that climate change issues disproportionally affect the vulnerable groups in society, entails that social work practitioners and educators need to be trained to combat the effects of climate change. Yet, social work curricula currently neglect the teachings of environmental concerns even though society is changing and requiring the profession to develop alongside it (Rambaree, 2020). Learning about climate change may enable social workers to appropriately engage in policies, respond to environmental challenges by mitigating or building resilience and become involved in discussions surrounding the topic.

Two key themes that can be incorporated within social work education are (1) the protection of the natural environment and (2) the means to engage with social, ecological and human development through cultural sensitivity. In addition, making use of existing approaches such as empowerment, anti-oppressive perspectives and a socially just value base can be foundational and environmentally responsive interventions (Rambaree, 2020). This requires integrating environmentally conscious concepts and ecological justice into curricula to address practice needs that focuses on the protection and empowerment of vulnerable populations in society.

In practice, social workers are equipped with the knowledge, values and skills to instil social change and are arguably both professionally and morally obligated to create mechanisms and conditions for a better environment for all (Rambaree, 2020). Given that the profession highlights the importance of understanding people within their environmental contexts, it is not farfetched to assume that social workers are expected to take on a role in tackling environmental threats to human wellbeing from an environmentally informed practice.

Schmitz et al. (2012) argue that fostering an environmentally sustainable future is most effectively achieved through an interdisciplinary approach. They suggest that social workers can play specific roles within an interdisciplinary approach to advance community sustainability. These roles encompass strengthening individual and community resilience, as well as facilitating practical interventions that encompass prevention, preparedness, relief, recovery and reconstruction of support services for those particularly vulnerable to the impacts of climate change (Dominelli et al., 2019). Recognising the disproportionate impact of climate change on vulnerable groups, an interdisciplinary approach becomes crucial for social workers in mobilising various stakeholders, such as the municipalities and environmental and climate disaster specialists (Närhi & Matthies, 2016; Walker, 2012). This approach facilitates collaborative problem-solving and knowledge-sharing in addressing climate change issues through an engaged and empowering process. The participation of community members alongside key stakeholders is essential for fostering positive change (Dominelli, 2012).

Social workers can play a pivotal role in both preventing and addressing the impacts of climate change by adopting various roles within the profession. The concept of intervening in climate change from a social work perspective involves leveraging existing ideals within the discipline that are familiar to social workers. For instance, social workers can repurpose roles such as case manager, mediator, social activism, community mobiliser and provider of psychosocial support to foster resilience and adaptation within communities (Dominelli et al., 2019).

In the capacity of a case manager, a social worker's primary responsibility is to assess the needs of client systems and coordinate the delivery of necessary services, often collaborating with other service providers. Acting as the interface between the client and their social environment, a case manager oversees the coordination of available support structures (Hepworth et al., 2010). In the context of climate change, this role involves linking communities with resources that align with their specific needs, aiming to improve their immediate natural environment and, consequently, their overall quality of life (Dominelli, 2012). A critical aspect of this role is connecting individuals with resources that enable them to access relief services after a catastrophic event.

On a broader scale, the active political involvement and social activism of social workers are imperative for environmentally informed practices (Mulvale, 2017). This role entails advocating for various preventive measures and policy changes that promote access to green technologies, fair distribution of resources, increased aid and the addressing of climate change on both international and national fronts (Dominelli, 2012). In the pursuit of enhancing both the natural world and human well-being, psychosocial support becomes particularly crucial following natural disasters (Dominelli et al., 2019). Social workers can address individual stress stemming from climate change through counselling. The potential catastrophes affecting the physical, economic, social, cultural or spiritual aspects necessitate social workers to assist communities in coping with the aftermath of natural disasters (Dominelli, 2012; Dominelli et al.,

2019). For extended psychosocial support, social workers can conduct educational campaigns aimed at behavioural modification and consciousness raising, both at the individual and community levels to promote sustainable practices.

Although there is limited research on climate change in South Africa, findings by Nel (2019) conclude that social workers in Tshwane identified that ideals related to environmental social work frameworks were not included within practice because of lack of knowledge of different intervention approaches, thus necessitating research that links social work and climate change from a South African context. What can be noted is that social work assessments and reimagining of existing roles are fundamental in disaster interventions to identify the level of risk and need after a natural disaster, as has been observed in recent years. However, a holistic and integrated approach to advance these interventions in South Africa still requires empirical research (Shokane, 2019), to foreground an understanding of the effects of climate change on vulnerable populations.

Conclusion and Recommendations

In conclusion, climate change as a crisis is a global phenomenon that necessitates a combined effort which is interdisciplinary and intercontinental for any societal positive outcomes. For the social work profession particularly, it is important to focus on those most vulnerable in society because their livelihoods are at risk of being greatly impacted by climate change. Given that existing social exclusion, marginalisation and social injustices will intensify because of climate change, social workers have an important role to play in reducing carbon emissions and promoting clean energy, but to also protect the vulnerable populations from the negative impacts of climate change, especially women and those living in poverty. This directly correlates with the profession as being charged with enhancing human wellbeing.

By making use of the social justice and environmental justice position, social workers in South Africa can embrace working towards and advocating for human rights in respect to environmental concerns. In doing so, a holistic understanding of how to respond to the impacts of climate change can be fostered by the profession. However, for social workers to have a comprehensive understanding and response, the effects of climate change upon vulnerable groups need to be better understood and considered. Given that climate change can exacerbate existing inequalities amongst women and people living in poverty, it is important to discuss in the context of linking social work to climate change. In terms of coping strategies for those living in poverty, one needs to first tackle the existing structures in society that perpetuate poverty and inequalities. However, building on resilience and adaptation may also empower communities to develop their own solutions, by working on sustainable development practices means those living in poverty can be simultaneously assisted in relation to climate threats.

Although the empowerment of women is part of South Africa's stride towards sustainable development, and some progress has been made towards gender equality, it is against this backdrop why women are a vulnerable population when considering the effects of climate change. Thus, climate change matters to vulnerable populations within society because it will disproportionally exacerbate the already existing injustices and marginalisation of the poor and women. Hence, social workers have a role to shape contemporary responses to societal challenges such as climate change guided by the principles of social justice and collective responsibility.

References

Almer, C., & Winkler, R. (2017). Analyzing the effectiveness of international environmental policies: The case of the Kyoto Protocol. *Journal of Environmental Economics and Management, 82*, 125–151. https://doi.org/10.1016/j.jeem.2016.11.003

Alston, M. (2013). Environmental social work: Accounting for gender in climate disasters. *Australian Social Work, 66*(2), 218–233. https://doi.org/10.1080/0312407X.2012.738366

Boddy, J., & Hetherington, T. (2013). Ecosocial work with marginalized populations: Time for action on climate change. In M. Gray, J. Coates, & T. Hetherington (Eds.), *Environmental social work* (pp. 46–61). Routledge.

Bowles, W., Boetto, H., Jones, P., & McKinnon, J. (2018). Is social work really greening? Exploring the place of sustainability and environment in social work codes of ethics. *International Social Work, 61*(4), 503–517. https://doi.org/10.1177/0020872816651695

MF Chersich CY Wright F Venter H Rees F Scorgie B Erasmus 2018 Impacts of climate change on health and wellbeing in South Africa International Journal of Environmental Research and Public Health 15 1 14 https://doi.org/10.3390/ijerph15091884

J Coates M Gray 2012 The environmental and social work: An overview and introduction International Journal of Social Welfare 21 230 238

Department of Environmental Affairs. (2012). *National climate change response: White paper. Pretoria*. https://www.gov.za/sites/default/files/gcis_document/201409/nationalclimatechangeresponsewhitepaper0.pdf

Department of Environment, Forestry & Fisheries. (2019). *National climate change adaptation strategy*. https://www.environment.gov.za/sites/default/files/docs/nationalclimatechange_adaptationstrategy_ue10november2019.pdf

Department of Statistics South Africa. (2023). *Quarterly labour force survey*. https://www.statssa.gov.za/publications/P0211/P02111stQuarter2023.pdf

Diffenbaugh, N. S., & Burke, M. (2019). Global warming has increased global economic inequality. *Proceedings of the National Academy of Sciences, 116*(20), 9808–9813. https://doi.org/10.1073/pnas.1816020116

L Dominelli 2011 Climate change: Social workers' roles and contributions to policy debates and interventions International Journal of Social Welfare 20 430 438 https://doi.org/10.1111/j.1468-2397.2011.00795

L Dominelli 2012 Green social work: From environmental crises to environmental justice Polity Press

Dominelli, L., Nikku, B. R., & KU, H. (2019). Green social work in theory and practice. In *The routledge handbook of green social work*. Routledge.

Garland, R. M. (2014). National policy response to climate change in South Africa. *South African Medical Journal, 104*(8), 584–584. https://doi.org/10.7196/SAMJ.8605

Germain, C. B & Gitterman, A. (1980). *The Life Model of Social Work Practice*. Columbia University Press.

D Hepworth RH Rooney GD Rooney K Strom-Gottfried J Larsen 2010 Direct social work practice 8 Brooks/Cole

Hijioka, Y., Lin, E., Pereira, J.J., Corlett, R.T., Cui, X., Insarov, G.E., Lasco, R.D., Lindgren, E. & Surjan, A. (2014). Asia. In V. R. Barros, C. B. Field, D. J., Dokken, M. D. Mastrandrea, K. J. Mach, T. E. Bilir, M. Chatterjee, K. L. Ebi, Y. O. Estrada, R. C. Genova, B. Girma, E. S. Kissel, A. N. Levy, S. MacCracken, P. R. Mastrandrea & L. L. White (Eds.), *Climate Change 2014: Impacts, Adaptation, and Vulnerability*. Part B: Regional Aspects. Contribution of Working Group II to the Fifth Assessment Report of the Intergovernmental Panel on Climate Change (pp. 1327–1370). Cambridge University Press.

Housing Development Agency. (2012). *South Africa: Informal settlements status*. http://www.thehda.co.za/uploads/files/HDA_Informal_settlements_status_South_Africa.pdf

IFSW. (2014). *Global definition of social work*. https://www.ifsw.org/what-is-social-work/global-definition-of-social-work/

IFSW. (2022). *The role of social workers in advancing a new eco-social world.* https://www.ifsw.org/the-role-of-social-workers-in-advancing-a-new-eco-social-world/

International Union for Conservation of Nature. (2015). Gender and climate change: strengthening climate action by promoting gender equality. *Online*, https://iucn.org/sites/default/files/2022-07/gender_and_climate_change_issues_brief_cop21_04122015.pdf

SP Kemp 2011 Recentring environment in social work practice: Necessity, opportunity, challenge British Journal of Social Work 41 1198 1210 https://doi.org/10.1093/bjsw/bcr119

T Kompas VP Ha TN Che 2018 The effects of climate change on GDP by country and the global economic gains from complying with the Paris climate accord Earth's Future 6 1153 1173 https://doi.org/10.1029/2018EF000922

A Lombard JM Twikirize 2014 Promoting social and economic equality: Social workers' contribution to social justice and social development in South Africa and Uganda International Social Work 57 4 313 325 https://doi.org/10.1177/0020872814525813

Madzwamuse, M. (2010). *Climate governance in Africa: Adaptation strategies and institutions.* Heinrich Böll Foundation. https://wrd.unwomen.org/practice/resources/climate-governance-africa-adaptation-strategies-and-institutions

MA Masoga SL Shokane 2019 Viewpoint: Indigenous knowledge systems and environmental social work education: Towards environmental sustainability Southern African Journal of Environmental Education 35 1 11 https://doi.org/10.4314/sajee.v35i1.14

McCloy, J. (2020). Kyoto Protocol vs Paris Agreement: Key Differences to Know Kyoto Protocol vs Paris Agreement: Key Differences to Know, *Online*, https://greencoast.org/kyoto-protocol-vs-paris-agreement/

K Morgaine M Capous-Desyllas 2020 Anti-oppressive social work practice: Putting theory into action 2 California State University

Muchacha, M. (2019). The gender dynamics of climate change on rural women's agro-based livelihoods and food security in rural Zimbabwe: Implications for green social work. *Critical and Radical Social Work, 7*(1), 59–72. https://doi.org/10.1332/204986019X15491042559655

JP Mulvale 2017 Reclaiming and reconstituting our understanding of "environment" in social work theory Canadian Social Work Review 34 2 169 186 https://doi.org/10.7202/1042887ar

Närhi, K., & Matthies, A. (2016). Conceptual and historical analysis of ecological social work. In J. McKinnon & M. Alston (Eds.), *Ecological social work: Towards sustainability* (pp. 1–98). Palgrave.

S Nel 2019 The role of social worker in promoting environmental justice for sustainable communities form NGOs' perspective in Tshwane. Unpublished master's thesis University of Pretoria

Niang, I., Ruppel, O. C., Abdrabo, M. A., Essel, A., Lennard, C., Padgham, J. & Urquhart, P. (2014). Asia. In V. R. Barros, C. B. Field, D. J. Dokken, M. D. Mastrandrea, K. J. Mach, T. E. Bilir, M. Chatterjee, K. L. Ebi, Y. O. Estrada, R. C. Genova, B. Girma, E. S. Kissel, A. N. Levy, S. MacCracken, P. R. Mastrandrea, & L. L. White (Eds.), Africa. In *Climate change 2014: Impacts, adaptation, and vulnerability. Part B: Regional Aspects. Contribution of working Group II to the fifth assessment report of the intergovernmental panel on climate change* (pp. 1199–1265). Cambridge University Press.

Norton, C. L. (2012). Social work and the environment: An ecosocial approach. *International Journal of social welfare, 21*(3), 299–308. https://doi.org/10.1111/ijsw.2012.21.issue-3, https://doi.org/10.1111/j.1468-2397.2011.00853.x

Patel, L., & Hochfeld, T. (2013). Developmental social work in South Africa: Translating policy into practice. *International Social Work, 56*(5), 690–704. https://doi.org/10.1177/0020872812444481

R Pierrehumbert 2019 There is no plan b for dealing with the climate crisis Bulletin of the Atomic Scientists 75 5 215 221 https://doi.org/10.1080/00963402.2019.1654255

Pinto, I., Zachariah, M., Wolski, P., Landman, S., Phakula, V., Maluleke, W., Bopape, M., Engelbrecht, C., Jack, C., McClure, A., Bonnet, R., Vautard. R., Philip, S., Kew, S., Heinrich D., Vahlberg, M., Singh, R., Arrighi, J., Thalheimer, L., van Aalst, M., Li, S., Sun, J., Vecchi, G., Yang, W., Tradowsky, J., Otto, F. E. L., & Dipura, R. (2022). Climate change exacerbated rainfall causing devastating flooding in Eastern South Africa. *World weather attribution,* 1–39. https://doi.org/10.25561/103482

K Rambaree 2020 Environmental social work: Implications for accelerating the implementation of sustainable development in social work curricula Journal of Sustainability in Higher Education 21 3 557 554 https://doi.org/10.1108/IJSHE-09-2019-0270

Schmitz, C. L., Matyók, T., Sloan, L. M., & James, C. (2012). The relationship between social work and environmental sustainability: Implication for interdisciplinary practice. *International Journal of Social Welfare, 21.*

Shokane, A. L. (2019). Social work assessment of climate change: Case of disasters in greater Tzaneen municipality. *Jàmbá: Journal of Disaster Risk Studies, 11*(3), 710. https://doi.org/10.4102/jamba.v11i3.710

Statistics South Africa. (2022a). *Gender series volume IX: Women empowerment, 2017–2022.* Statistics South Africa.

Statistics South Africa. (2022b). *Statistical release p0301.4.* Statistics South Africa.

Statistics South Africa. (2023a). *Consumer price index*, April 2023. Statistics South Africa.

Statistics South Africa. (2023b). *Quarterly labour force survey: Quarter 1.* Statistics South Africa.

M Strydom U Schiller J Orme 2020 The current landscape of child protection services in South Africa: A systematic review Social Work/maatskaplike Werk 56 4 383 402

Teater, B. (2014). Social work practice from an ecological perspective. In: C. W. LeCroy (Ed.), *Case studies in social work practice* (3rd ed.). Brooks/Cole.

UN Women and United Nations Department of Economic and Social Affairs. (2022). *Progress on the sustainable development goals: The gender snapshot.* https://data.unwomen.org/publications/progress-sustainable-development-goals-gender-snapshot-2022

Valodia, I. (2023). *South Africa can't crack the inequality curse. Why, and what can be done*? October 9, 2023. The Conversation AFRICA.

Walker, G. (2012). Climate change oppression: Media production as the practice of freedom. *Consilience: The Journal of Sustainable Development, 9*(1), 97–106. https://www.jstor.org/stable/26476128

CHAPTER 12

The Power of Small-Scale Interventions in Progressing Climate Adaptation and Resilience in the Caribbean: A Case Study of Trinidad and Tobago

Lauren Bain and Sara Low

The Climate Reality of Small Island Developing States (SIDS)

Small Island Developing States (SIDS) are projected to face up to US$26 billion in damages annually by 2050, which represents approximately 5% of their combined GDP if climate change remains unmitigated (United Nations, 2024a). These states, which include 39 nations worldwide, many in the Caribbean, are characterized by their reliance on external

L. Bain (✉)
Nature Positive, London, UK
e-mail: laurenbain@outlook.com

S. Low
EY Caribbean, Port of Spain, Trinidad and Tobago

D. D. Joseph et al. (eds.), *Climate Change*,
https://doi.org/10.1007/978-3-031-83037-2_12

markets, limited internal resources, and fragile land and marine ecosystems (United Nations, 2024a). This unique combination makes them particularly vulnerable to global risks, especially climate change.

Despite ongoing global efforts, progress toward climate mitigation has been slow, and the pace of climate change is outstripping many countries' ability to adapt. For SIDS, which stand at the forefront of climate impacts, resilience is more critical than ever. The United Nations Framework Convention on Climate Change (UNFCCC, 2021) defines climate resilience as "having the capacity to prevent, anticipate, and absorb climate extremes and slow-onset events, as well as adapt and transform development pathways in the longer term." While governments play a significant role in addressing these challenges, they cannot protect against the full spectrum of impending climate realities on their own. This is especially true in SIDS, where government budgets are stretched thin and allocated to other socio-economic issues. In fact, Caribbean SIDS have some of the highest global debt rates, which reached 70.5% of their GDP in 2018 (International Institute for Sustainable Development, 2019). This further limits their ability to respond effectively to climate shocks. While governments typically bear the responsibility of mitigating large-scale impacts, it is clear that citizens also have a role to play. As Mees et al. (2019) emphasize, "it is imperative that we do not wait for the government alone to act." In the face of climate change, the role of the public, alongside government efforts, is essential.

Evolving Roles of Governments and Citizens in Climate Resilience

Historically, governments have protected the social welfare of their populations through "social contracts" agreements that outline the responsibilities and mutual benefits between civil society and the state (O'Brien et al., 2009). These contracts commit the state to providing services like law and order, healthcare, and education, while citizens contribute by paying taxes, voting, and abiding by laws. However, traditional social contracts often neglect protections against the urgent risks posed by anthropogenic climate change, exacerbating disparities in the distribution of power and responsibility. With climate-related risks occupying five of the top ten positions among the most severe global risks over the next decade (World

Economic Forum, 2024), it is evident that SIDS current power structures are insufficient to address the extreme weather, biodiversity loss, and resource shortages that are accelerating as climate change worsens.

Ideally, social contracts should represent the interests of all stakeholders. In practice, however, they often reflect societal inequities, enabling those with power to shape social relations to their advantage (Blackburn & Pelling, 2018). This disparity is particularly pronounced among vulnerable groups such as women, Indigenous peoples, and persons with disabilities who often lack the platform to contest the status quo (O'Brien et al., 2009). Addressing these inequities is critical, as is ensuring the protection of ecosystems and the species they support. Tackling climate resilience requires targeted efforts that are fair, just, and inclusive, with decision-making that accounts for all parties. Reimagining the social contract in this context challenges the traditional foundations of economic growth and prosperity. It also forces us to reconsider the state's role in providing protection in a world shaped by global environmental challenges.

The Context for Resilience in Trinidad and Tobago

Trinidad and Tobago (T&T), a multicultural twin-island nation in the Caribbean, has a population of approximately 1.4 million people (Central Statistical Office, 2024). The country's economy relies heavily on resource-intensive industries, particularly oil and natural gas, as well as agriculture, manufacturing, tourism, and financial services (World Bank, 2024). These strong economic sectors have enabled the government to subsidize essential services like water and electricity. However, in the face of natural gas shortages and tightening national budgets, the country is under increasing strain. Communities that depend on low living costs for survival are feeling the pressure. Beyond economic challenges, Trinidad and Tobago, like many Caribbean nations, grapple with systemic issues stemming from colonialism, racial inequalities, and a capitalist, extractive economic model. These legacies shape the distribution of resources and influence human–environment relationships, making climate resilience a complex issue that disproportionately affects the most vulnerable groups.

Though direct efforts toward consideration for the most vulnerable of groups are not evident, The Government of Trinidad and Tobago has made clear climate commitments to reducing overall greenhouse gas emissions by 15%, as outlined in its Nationally Determined Contribution (NDC) to the United Nations Framework Convention on Climate Change (UNFCCC). However, 73% of these emissions come from the industrial sector, a cornerstone of the national economy. Achieving this target requires a delicate balance between building climate resilience and maintaining economic growth.

To address climate challenges, the government has incorporated climate considerations into several policies, including the National Climate Change Policy, the National Development Strategy, the Integrated Water Resource Management Policy, and the Integrated Coastal Zone Management Policy. While these initiatives reflect the country's growing commitment to climate action, projections from the Government of Trinidad and Tobago (2021) suggest that without effective adaptation, the nation's GDP could decrease by up to 30% by 2050, and as much as 82% by 2100. These figures highlight the urgent need for robust adaptation strategies.

The escalating costs of climate impacts exceed the capacity of many countries, including Trinidad and Tobago, to respond effectively. As such, the government must recognize the need to collaborate with civil society organizations. Trinidad and Tobago's National Adaptation Plan (GORTT, 2021a) notes that "effective adaptation and long-term resilience-building depend on the meaningful collaboration and participation of academia, research institutions, public and private sectors, non-governmental organizations (NGOs), community-based organizations (CBOs), business and industry organizations, and the citizenry at large." While governments bear the responsibility for large-scale impacts, citizens, too, must contribute to climate resilience efforts (Mees et al., 2019).

The Role of Small Actors in Climate Action in Trinidad and Tobago

Small to medium-sized enterprises (SMEs) are a vital part of Trinidad and Tobago's economy, accounting for more than 30% of GDP and 80% of registered businesses (International Labor Organization, 2023). Despite this, government policies on sustainability for SMEs, and micro, small, and medium enterprises (MSMEs), (businesses with a revenue between $250,000 and $8 million T&T dollars per annum) (Ministry of Youth

Development and National Service (MYDNS, 2024) remain underdeveloped, often due to the misconception that these businesses lack interest in climate action and lack the breadth to cause significant impact. While individually there is merit to this argument, collectively there are over 20,000 MSMEs operating in Trinidad and Tobago (International Labor Organization, 2023), indicating that this is a significant segment of the business sector whose contribution toward both climate change and resilience should not be downplayed.

MSMEs play a crucial yet often overlooked role in climate action, as they are seen as a 'hard to reach' group that lags behind larger businesses and the public sector in relation to climate action (Hampton et al, 2022). Additionally, MSMEs face a wide breadth of challenges in building climate resilience which involves limited access to resources (including data and information relating to adaptation options, finance, technical skills, and capacity), lack of skills related to business development and start-up skills, and lack of an enabling environment for climate action (Montmasson-Clair, et al., 2019). That said, there are also a number of solutions that have been studied ranging from addressing policies, to developing capacity, infrastructure, access to relevant data, technology, building markets (Montmasson-Clair et al., 2019) that will allow MSMEs to improve their climate resilience through building business models on climate-focused goods and services. However there are various levels of support required by the state, private sector, and financial institutions, including inclusive policy development and consultation processes in the public sector, and the development of procurement processes and relationships with MSMEs by large private sector players to encourage investments into adaptation.

Thus, while there are few institutional facilities to support them, many MSMEs in Trinidad and Tobago take a values-based approach to sustainability, driven by their own principles rather than government mandates (Hampton et al., 2022). This localized understanding of climate impacts enables them to address sustainability in ways that align with both business interests and community needs. MSMEs play an important role in promoting collective climate action, often providing a bridge between awareness and practical decision-making. As Plummer et al. (2022) point out, climate change can feel abstract for many, especially when solutions are implemented without adequate consultation to relevant stakeholders. MSMEs, with their deep roots in the community, are better positioned to address the everyday realities of climate change, helping to close the gap between policy and practice. This holds true for non-governmental organizations (NGOs) as well, which are any non-profit, voluntary citizens' group which is organized on a local, national, or international level" (United Nations, 2024b), as these organizations, too, sit at the intersections of community needs and organizational capacity.

Community-focused MSMEs can also help bridge the gap between climate awareness and decision-making, promoting collective action. Plummer et al. (2022) highlight that despite increasing awareness, climate change remains abstract for many and is not embedded in everyday habits and routines. This disconnect is often exacerbated by top-down government approaches, which, without proper consultation, can provide a false sense of security and fail to consider the nuances of power relations and local adaptation practices (Simon et al., 2019). Simon et al. (2019) argue that "adaptation should be seen as part of the dynamics of societies rather than simply being a technical adjustment to biophysical change by society," emphasizing the need for localized, case-by-case approaches.

This challenge extends beyond local governments to global institutions and Western governments, which often impose development projects on the Caribbean. These initiatives are typically framed as grassroots but can operate under the assumption that local Caribbean modes of living and thinking are static or primitive, hindering development (Gahman et al., 2021). Plummer et al. (2022) call for a more critical, affective, and ethical approach, such as using emotional appeals through art to inspire climate action. In this context, the inclusion of MSMEs and community-led initiatives in climate action is essential. Given their significant economic role and ability to innovate at the local level, MSMEs can support national climate goals while fostering resilience in their communities. This chapter explores how MSMEs and NGOs contribute to climate resilience in Trinidad and Tobago, while also identifying the barriers that limit their broader impact.

Methodology

Utilizing a qualitative case study approach to understand how the chosen MSMEs and NGOs enhance Trinidad &Tobago's climate resilience efforts as well as to provide more context around the importance of locally focused, small-scale contributions to climate action. By conducting interviews with proprietors, we were able to obtain detailed primary data that captured the nuanced views and firsthand experiences of participants (Bryman, 2012), making them an excellent tool for understanding the dynamic climate adaptation and resilience ecosystem in Trinidad & Tobago. The aim was to gather insights directly from those actively involved in climate initiatives, ensuring the authenticity and relevance of the data obtained.

Organizations were selected against the inclusion criteria based on several key considerations. First, organizations had to fit the definitions of MSMEs and NGOs as provided in the study. This was necessary to maintain consistency in the sample, aligning with both local and global definitions of NGOs and MSMEs. Second, organizations needed to be fully operational in Trinidad or Tobago at the time of the interview to ensure that the responses reflected current conditions and the specific context of the local setting. Third, the mandate and projects of the organizations had to be related to environmental and/or social sustainability. Both direct and indirect mandates and projects were accepted to allow for a broad examination of what climate action encompasses, ensuring relevance to the research question. Finally, willingness to participate was a key criterion, as not all organizations were able or willing to share their stories within the specified timeframe. Therefore, priority was given to those organizations who expressed a willingness to meet with us and participate in the research. Purposive and snowball sampling (Bryman, 2012) was used due to the limited selection pool. Potential participants with relevant insights were approached via email and through expert recommendations in the local sustainable development field. Table one provides a full list of the basic demographics of the final selection of organizations. This selection process has noted limitations but was guided by the need to capture a diverse range of perspectives from key stakeholders in the local climate ecosystem (Table 12.1).

Data was collected through 40–60-minute semi-structured interviews conducted via Microsoft Teams. This format allowed for flexibility in exploring various themes while ensuring that all key areas of interest were covered. Interview questions were designed to collect detailed information about the participants' initiatives, challenges, successes, and perceptions regarding their role in climate adaptation and resilience. Prior to the interviews, participants were required to sign a consent form detailing their rights to anonymity, outlining the research process, and explaining how data would be used and stored. The consent process ensured that participants were fully informed and agreed to the terms

Table 12.1 Basic demographics of interviewees by authors

Respondent (Res) number	*Organization type*	*Location*	*Sector*
1	MSME	Trinidad	Sustainable agriculture
2	NGO	Trinidad	Community climate justice
3	MSME	Trinidad	Private sector/ green product innovation
4	NGO	Trinidad	Caribbean climate justice
5	NGO	Trinidad	Environment and natural resources
6	NGO	Tobago	Ecology, oceans and fisheries
7	MSME	Tobago	Circular economy

of their involvement in the study, which is voluntary. Data was securely stored and accessible only to the research team. Transcripts will be deleted after five years.

Inductive thematic analysis was utilized to analyze the interview data. This method was chosen for its effectiveness in identifying, analyzing, and reporting patterns within data "without trying to fit it into a pre-existing coding frame" (Braun & Clarke, 2006). Throughout analysis of interview transcripts, emerging patterns in the data were coded systematically and collated into potential themes. These themes were then formally defined to refine the specifics of each, allowing them to be woven into the discussion and linked back to the literature reviewed to create the structure of the case study.

Researcher's Reflexivity

This research is deeply personal to the authors as two sustainability professionals who grew up in Trinidad and have spent over eight years working, researching, and advocating for climate action. Our experience in the climate sector has shown us the resilience of communities and their ability to adapt to climate issues, both where the support of local MSMEs and NGOs was available, and even where it was not. While we are connected to the private sector, we recognize that many talented individuals within

MSMEs and NGOs are often overlooked—not due to a lack of innovative ideas—but because of insufficient resources and support. This study aims to highlight the critical role of these small-scale actors in building T&T's climate resilience, illustrating how their work is not only relevant but essential. It is a call to uplift these emerging climate leaders and create a future where tackling climate challenges is seen as a noble and rewarding pursuit. We are driven by the belief that with the right support, these small-scale actors can become powerful agents of change, turning innovative ideas into practical solutions for a more resilient T&T, and hope to inspire greater recognition of the value of local solutions in creating a sustainable future.

Results

During the analysis of interview transcripts, several themes and subthemes emerged to paint a picture of the current state of climate action and resilience in T&T. As part of the analysis process, similar ideas were grouped under umbrella themes. Five key themes emerged across the interviews: (1) the importance of communicating climate science in layman's terms, (2) a resistance to change in T&T which can impact MSMEs and NGOs from implementing new projects in the climate space, (3) the link between community and climate resilience, (4) the disconnect between governments and communities, and (5) Factors influencing the effectiveness of small-scale interventions. These themes are explored in this section using direct quotations from respondent interviews (included in italics for ease of reference) and reference to existing literature in the space.

Theme One: Communication of Climate Science

Immediately, the inaccessibility of climate information across T&T became clear, and in some cases, a lack of engagement with knowledge-sharing attempts due to the fact that the issue feels completely separate from people's daily lives. As one respondent put it,

> nobody explains...climate change in a way that [communities] would understand. (Res 6)

Other respondents involved in climate-related projects echoed this sentiment, speaking to the *"lack of understanding as to why we're doing what we're doing" (Res 1)* and the fact that often, *"stakeholders have no clue what climate change is" (Res 6)*.

In addition to lack of technical education on climate matters, there is also the issue of livelihoods being a priority. When the *"most important thing is daily income" (Res 6)*, especially in cases where there is not sufficient information available for people to learn about how climate change relates to their lives, survival will trump climate action every time. This is aligned with the idea that in the Caribbean, socio-economic conditions, and structures dominate the public consciousness (Plummer et al., 2022) and can act as a barrier toward the adoption of a climate conscious mentality, especially if this is seen to be altering one's ability to provide economically.

One point of view was particularly interesting, and this spoke to the idea that in climate space, practitioners may *"take for granted that everybody knows what we're talking about" (Res 4)*. Given the context that engagement with climate topics already has barriers, this quote speaks to the need for innovation and change in the way that we execute such communication. The idea that learning around adaptation is a social process (Simon et al., 2019) that should be community centric comes to the forefront here, and integrating other methods of communication and storytelling, such as arts and theatre, may open the door to new types of knowledge sharing across *"different audiences, different age groups and spaces" (Res 4)*.

Theme Two: Resistance to Change

The second heavily recurrent theme was the resistance to change that is seen across society in T&T. Although T&T's National Adaptation Plan (GORTT, 2021a) and National Communication (GORTT, 2021b) refer to the importance of both mitigation and adaptation for the prosperity of our country, it was perceived by a respondent that there is a disconnect between "*where we want to be and the cultural mindset of not wanting to try new things" (Res 7)*.

Another respondent noted that, even outside of the climate space,

> people are ascribed to what they know because they know it, but not because it works. (Res 4)

In any sector, this will prevent innovation and progress, and however, in the climate space it can be even more detrimental as it widens the gap between climate change as a concept and climate action (Plummer et al., 2022) which can result in a version of "brain drain" in which innovative climate projects are forced to seek support and funding abroad. This was noted in interviews, as respondents consistently noted that they received *"more recognition" (Res 3) and "easier success" (Res 7)* when interacting with stakeholders outside of T&T related to their work. Even when mutually beneficial partnerships could be formed to create an ecosystem of positive benefits for both parties locally, those seem not to be met with the enthusiasm that it is with foreign entities who have a much better appetite to forge partnerships and relationships. This trend is so much so that local MSMEs, while intentions were to change the local landscape, have focused their efforts outside of T&T where they gain more traction, rather than on the local scale where people "*don't really appreciate what [they] do" (Res 1).*

It was highlighted that the only way to address this resistance is through behavioral change, which starts with education for communities to understand "*why [climate solutions are] important and why it's necessary*" *(Res 2)*. One respondent mentioned the challenge with actioning innovative ideas locally is also that many stakeholders are resistant to taking risks and want to see proof of success first before implementing, since "*very few people ever take on new or unchartered territory" (Res 4).*

Stakeholders also need education to resolve their concerns of "*why should I spend money to green something?" (Res 2) which* demonstrates resistance to taking positive action based on existing beliefs. Beyond money, another factor in resisting positive change is lack of understanding of the benefits to the community. For example, one respondent recalled that *"for years, everybody in our community thought [we] were crazy..." (Res 1) for* going against the status quo of their industry and implementing sustainable practices. It was only over time and with active education of the community that behaviors shifted, and culture changed to embrace the new practices. However, even when culture is changed, other challenges that small organizations face is that stakeholders don't allocate appropriate resources to support the projects. That's where partnerships across other small entities and NGOs are critical. When small organizations can partner with organizations that are aligned in their goals, their impact is extended.

Theme Three: The Role of Small Actors in Maintaining Climate Resilience

> Small-scale initiatives are really building resilience at the local level and that's what actually matters...people who are on the front lines facing the impacts on their livelihoods and their world (Res 5).

As identified by Mees et al. (2019), it is not only within the government's remit to build resilience against climate change. Civil society also has an important role in filling the gaps that the government cannot. Currently, MSMEs and NGOs in the climate space understand the importance of their work toward building resilience in the wider community, and these entities have strong foundations of trust and understanding of how best to serve these communities given their specific local context, merging local and traditional knowledge with scientific concepts to contribute to adaptation and resilience solutions. With this kind of leverage, one respondent, who operates a permaculture farm in their local community, described their purpose as being a *"living example of what could be" (Res 1)* if communities were able to harness the benefits of these practices. They work toward achieving this goal through knowledge-sharing sessions with a "pay what you can" model for farmers living in the area.

These types of community-led, small-scale interventions were common across respondents from both MSME and NGO groups, with many noting that climate resilience is directly enhanced when communities are able to be financially independent and access the basic resources, they need to survive in the face of the climate crisis. In two of the business models of the MSMEs interviewed, integrating community-level job opportunities for those who need access to income was key. By creating this reciprocal relationship of providing employment to vulnerable groups and accessing reliable skill sets to accomplish tasks, these business models are supporting climate resilience at the grassroots level from both a financial and a solutions-based standpoint. One respondent described their solution as not providing the resource the community needed, but providing the money that the community would need to gain better and sustained access to that resource:

> I'm not giving you water, but I am giving you more money in your pocket. (Res 3)

This sustained financial independence, while indirectly related to climate change, is an important piece of the resilience puzzle that should not be overlooked. As one NGO rightfully put it:

> the conversation about climate change is not, in fact, about climate change at all, but just about the humanity of it. (Res 4)

However, interviewees also recognized that "*no one entity can solve all the problems*" *(Res 2)* alone. One respondent criticized the global approach to climate change saying, "*we haven't treated [climate action] as a community effort*" *(Res 3)*. It is commonplace for entities, governments, and people to work in silos, when in reality, the most effective solutions come from "*coalitions*" *(Res 7) between* different organizations, allowing for the sharing of resources toward a similar goal. When collaboration is prioritized, MSMEs and NGOs have the ability to join with larger organizations, extending their reach and allowing them not to *"feel so small"* *(Res 7)*. One respondent shared their aim to build this type of partnership ecosystem where the services of one support and attract customers for other service providers, thus creating a *"collective experience"* *(Res 5) and* attracting higher volumes of revenue through collaborative efforts. Other respondents mentioned that many of their projects have been *"achieved through collaboration"* *(Res 2)*, or they *"tried to support each other as much as they could"* *(Res 1)* with the resources that they were able to access, even when these were limited. In these ways, the role of civil society is filling the gaps that are hard for policy and government intervention to reach due to their localized and specific contexts and bringing people along on the journey to shared community resilience.

Theme Four: The Disconnect Between Government and Communities

The idea of naturally occurring and differentiated roles of government and communities in addressing climate change mitigation and adaptation (Mees et al., 2019) was also noted as a feature in T&T's climate space. These roles contribute to a disconnect, whether real or perceived, between government action and community needs which then cascades into government projects and action. For example, the reality of T&T being a hydrocarbon economy that needs to focus on mitigation in policy and strategy was raised. However, when government resources are only focused on one area, there leaves space for a perception that there hasn't been as *"much of a focus on resilience and adaptation" (Res 5)* for other sectors and communities. One respondent expands on this, referencing the fact that smaller organizations and community-led groups may feel left out of the national narrative and unsupported or not prioritized in government projects due to the focus not considering them -

> small marginalized groups that are not really, I think, effectively supported by these big national projects and big regional projects that are focused more on things like infrastructure. (Res 5)

The perceived lack of trust in government initiatives to meet the needs of smaller climate projects became even more apparent when one respondent remarked that when it comes to these projects, *"there are no emergencies" (Res 4)*. Another echoed this sentiment, stating that *"it comes down to who you know" (Res 7)*. These sentiments are reminiscent of the idea that top-down approaches are often implemented without proper procedures or processes to ensure that adequate attention is placed on critical issues. In fact, despite the recognized importance of grassroots and community-level work to complement the gaps in reach of government (Mees et al., 2019), those working with NGOs and SMEs recognize that government organizations and the private sector prefer to collaborate with larger organizations that have access to "*more money because… they might be able to see results faster" (Res 7)*. However, a missing puzzle piece is that the community-led and grassroots initiatives often have the knowledge and context of the issues and sit in "*a front row seat*" *(Res 4)* where they may be best equipped to tackle the impacts of climate change at their local level.

There is also a sentiment that the government has a tendency of thrusting a project upon a community, rather than consulting and engaging them in a meaningful way to understand their needs. Additionally, when considering that no populations are homogenous, especially amid T&T's melting pot of cultures, projects must be sensitive to the diverse identities of the people they serve. However, NGOs and MSMEs sentiments toward consultations, when done at all, are that they are often done as a check-box exercise, rather than to genuinely hear the concerns of the community, whereas *"stakeholder engagement only makes sense when it's informed and meaningful"* *(Res 6)*. Often, community members do not attend these consultations because *"there is a level of mistrust with the government"* *(Res 2)* and as a result, communities often *"do not feel represented"* *(Res 6)* by the people making decisions on their behalf. To further support this point, respondents noted that there have been cases where communities object to government-led projects, but projects ensue regardless.

To overcome these challenges, MSMEs and NGOs highlighted there needs to be more facilitation and support from the government for sustained climate action and related business. In fact, one MSME in the Carnival industry called for the reallocation of the national budget toward their industry to be further broken down to incorporate resource allocation for the sustainable development of that sector. This would simultaneously contribute to reducing the resistance to change since it would provide opportunities instead. This moves beyond grant money, highlighting a need for long-term financing that does not cover directly climate-related projects only, but also projects that can facilitate climate projects indirectly. For example, financing that facilitates the development of infrastructure to support future climate work, such as support for the building of facilities for MSMEs and NGOs that attract eco-tourists to the region, or the reallocation of funds and development of supporting policies to promote innovative projects created by young people and new graduates. This policy and resource realignment can directly channel innovation and brainpower into solving the direst climate solutions we face today.

Theme Five: Factors Influencing the Effectiveness of Small-Scale Interventions

Montmasson-Clair et al. (2019) outlined the challenges of low- and middle-income countries in climate resilience work, including poor access to relevant data and guidance for adaptation options, socio-cultural barriers, and finance and capacity limitations. Though T&T is considered a high-income country, its national circumstances being a SIDS also lend itself to many similar issues.

> Politics is still the most important stumbling block and hindrance to sustainable development that we have. (Res 6)

The major theme raised by respondents as barriers to the effectiveness of their work was a lack of support that covered a number of sub-areas. This includes buy-in from the private sector, inaccessibility of financial support, political disinterest, lack of capacity to execute, and socio-cultural resistance. The combination of these challenges makes it difficult for climate entrepreneurs and innovators to hold space and make financially sustainable progress in the climate space in T&T. One respondent commented that "*this market may not be ready for this particular thing… my market may be somewhere else*"*(Res 3)*. Allowing this mindset to permeate into the climate space results in those who have the will and expertise to enact change leaving the country and to develop their ideas in countries that have better-suited enabling environments for success and larger appetites for innovation.

The financing challenge has multiple layers to it, the first being that financing is often highly inaccessible to small-scale initiatives. Global climate finance is often most available to large-scale, incorporated organizations with audited financial statements and other governance requirements. When looking at the local community level, these requirements and structure for fund allocation is not always feasible. Respondents also emphasized the need for "*getting the funds to those that need it most*" *(Res 5)* and the importance of intermediaries that can translate large-scale funding into "bite sized" funding programs that are more accessible. These intermediaries should not be the large banks, but rather those that have more roots and connections to the community level. Another hurdle with large-scale financing is the time requirements for distributing capital. One respondent noted that it took them "*several years to actually kick off the project [with the Green Fund (T&T's National Environmental Fund]*" *(Res 2)* due to significant delays and bureaucracy in the process. Years of delay on a project is enough to hamper and diminish enthusiasm and progress toward the end goal, as people need to be able to feed their families and cannot wait endlessly for project implementation.

Another issue raised is that larger entities that may have access to financing for small-scale projects may not see value in them, as the success of small-scale projects is contingent on those doing the work and having enough funds to sustain their livelihoods. To respondents interviewed for this research, the sentiment was that "*sustainability is more of a side effect than the main mission*" *(Res 7)* of the business agenda, especially when it is expressed to MSMEs and NGOs that larger companies' environmental and sustainability responsibilities have already been fulfilled by paying into

Trinidad's Green Fund. This idea should be further explored in future iterations of research to obtain more details around the perception that businesses are unwilling to pay into the Green Fund while also sponsoring and investing in green projects. Further understanding and addressing this is an important socio-cultural hurdle to overcome. Though it would be easier to imagine a climate working world where small organizations didn't have to convince large corporations of the importance of their work, it's also critical to understand that partnerships and support from large corporations go a huge way in terms of access to networks, finance, weight and respect in the business community and society by being affiliated. Thus, though skepticism and resistance in support is common, partnerships with large corporations are instrumental in the scaling and reach of many community projects.

The dichotomy between T&T's hydrocarbon economy but vulnerability to climate change is interwoven throughout this paper, and this emerged once more in interviews in the form of frustration from respondents. The frustration centers on the fact that partnerships with large corporations are critical to achieve climate goals, yet many small organizations feel as though the *"giving back [of large corporations] is so minimal [compared to] all the damage that they do" (Res 1)*. This point is important that large private sector players take on a significant enough role in supporting the development of climate solutions, resilience, and adaptation in comparison with the large-scale detrimental impacts that many of their operations have on the environment. While many separate Corporate Social Responsibility and Environmental, Social, and Governance programs, it is vital to ensure that these programs meet the needs of the country that the company operates in. If not, the risk of ineffective action for the sake of compliance and marketing, as well as continued climate impacts that can affect the viability of long-term business operations, is increased.

Conclusion

The main findings highlight key issues that are prevalent in the climate space in T&T, including communication barriers, resistance to change, the critical role of small actors, the disconnect between government and communities, and financing issues impacting the ability of smaller players to implement much-needed climate projects and interventions.

The themes explored underscore the complex interplay between socio-economic conditions, cultural mindsets, and institutional structures that shape T&T's climate resilience efforts and emphasize the need for a reassessment of the climate ecosystem to address persistent issues, bridge gaps, and ultimately strengthen the country's collective capacity to achieve its climate goals and enhance resilience across its communities. This aligns with much of the literature previously explored throughout the chapter, echoing key sentiments that economic interests can take precedence over climate action, and reiterating the institutional barriers to effective implementation of climate initiatives. However, the discussion presented in this chapter offers a deeper understanding of how these dynamics manifest in T&T's context, and specifically how they are evolving from the scenarios that are presented in existing literature. As the country's vulnerability to adverse climate events grows, the social landscape and attitude toward climate action evolves alongside it. This change has necessitated collective action and resilience-building to adapt to climate change and is making it harder to disengage with climate issues regardless of technical knowledge or economic positioning, as impacts are felt across all of society.

While this study provides valuable insights into the role of smaller actors in climate adaptation and resilience in Trinidad and Tobago (T&T), several limitations should be acknowledged. First, the research was conducted within a constrained time frame, which limited the opportunity to identify and engage with a broader pool of participants. As a result, the sample may not fully represent the wide range of MSMEs and NGOs engaged in climate-related work. Additionally, the limited availability of many MSMEs and NGOs, due to the demands of their ongoing projects, presented challenges in scheduling interviews, which may have restricted the scope of perspectives included. Furthermore, the focus on purposive sampling, which targeted MSMEs and NGOs specifically involved in climate resilience, created a relatively small pool of eligible organizations. This limited sample size may affect the generalizability of the findings, as the study primarily reflects the experiences of those who were available and willing to participate within the research timeframe.

However, despite these constraints, the information gathered offers critical insights into the contributions of smaller actors in advancing climate goals in T&T. These findings contribute meaningfully to the limited body of literature on climate resilience in the Caribbean, particularly in terms of understanding the unique challenges and opportunities faced by smaller players. Moreover, this research provides a valuable foundation for further exploration in this area, highlighting the need for policy interventions that better support MSMEs and NGOs in their efforts to address climate challenges. Future research should build on these insights to determine the most effective ways to center small-scale actors in national climate strategies and policies, with detailed action plans and timelines to achieve this. A full range of detailed suggestions for future initiatives to strengthen T&T's climate action and resilience through MSME and NGO engagement can be found in the final section of this chapter: "the way forward."

The Way Forward

Two statements must guide future climate action in T&T:

1. The Caribbean is vulnerable to climate change impacts despite not causing the problem, emphasizing the need for adaptation and resilience measures.
2. NGOs and MSMEs greatly contribute to the creation and maintenance of national climate resilience through their close ties with local communities and ability to execute grassroots, localized initiatives that create change on a micro level, enhancing both social and environmental well-being.

It is this well-being that enables the country to meet its larger climate goals. Based on the research undertaken, there are issues to be overcome for both groups, and that these hurdles cannot be removed alone. We have therefore developed a series of actionable recommendations for the larger players in the private and public sector to allow them to provide the support needed for these important initiatives to create long-term behavioral and physical change in the climate space. These suggestions are outlined below.

1. **Enhance Opportunities for Collaboration from Both the Public and Private Sector to Increase Capacity of NGOs and MSMEs to Execute Climate Work**

The ecosystem of climate change stakeholders, from those impacted to those actively engaged in climate-related work, is established within T&T. The issue is that no enabling environment currently exists for like-minded organizations to collaborate and grow. Actionable items to achieve greater support and collaboration among this ecosystem include:

- Creating policies with corresponding implementation plans to encourage MSME participation in climate action initiatives
- Revamping existing climate and sustainability funds to cater to MSMEs and community-led organizations carrying out work that links indirectly to climate, for, e.g., social-related work enhancing community capacity and resilience
- Creating incubators for ideas and networking spaces for the private sector, NGOs and all players to come together to discuss and execute projects
- Supporting NGOs and MSMEs to access financing for climate projects by helping them to navigate grant applications, network, build partnerships, and expand their reach

2. **Alter Communication Methods and Strategies When it Comes to Climate-Related Information**

Communication is crucial for enabling climate action, however, the communication methods and channels chosen must resonate with the public to be effective. Essentially, climate-related information needs to be contextualized to make sense in the lives of the people receiving it, and messages should be tailored to relate directly to their daily experiences and concerns. Actionable items to enhance communication strategies include:

- Exploring innovative techniques like art, storytelling, and theater to convey climate messages in engaging and relatable ways
- Leveraging local artists to create performances, art pieces, and installations that depict the impacts of climate change and the importance of climate action.
- Funding public art projects and partnering with community schools and community centers for climate-themed events.

- Using storytelling methods, for example, through the creation of videos or podcasts featuring local activists, scientists, and community leaders sharing their experience, knowledge, and advice on climate action and resilience may also broaden the accessibility of climate information. This can be planned and executed via a communications strategy to maximize reach.

3. **Long-Term Education Initiatives Should be Prioritized and Funded and Planned for Properly over Time to Create Sustained Behavior Change**

Across many interviews, the idea that innovation in the climate space was discouraged due to an almost innate resistance to change came up. Across many interviewees, we heard this repeated issue of having more support internationally, or being considered strange for living life sustainably, or people not being willing to support an idea that goes against the "norm" of the way that we communicate and live today. Behavior change of this magnitude cannot happen overnight and requires thoughtful planning and integration of new concepts, including innovation, sustainability, and challenging the status quo into education systems. It is also worth noting that while the impacts of this may take longer to occur, they are worth planning for as they will change the mindset of an entire generation of future leaders, politicians, inventors, philanthropists, climate scientists, and more. Both the private and public sector can take the lead in pushing these narratives, as it will benefit them in the long term. Actionable items include:

- Supporting students even if the education system has not yet integrated these themes into curriculum, for example, through training or graduate programs focused on innovation and sustainability.
- Developing scholarship programs that funnel top talent directly into cutting-edge climate research that will benefit the country and company.
- Designing and running innovation competitions that encourage youth to get involved and learn.
- Holding open forums with young people to allow them to bring their ideas to leaders (both public and private sector), no matter how

"unrealistic," and help them to workshop those ideas into tangible solutions.

4. **Participatory Governance and Inclusive Consultations Should be at the Forefront of Climate Policy and Strategy Development**

Consultations need to be meaningful to be impactful, and this is not only at the government level but across all sectors of business who may be implementing a climate-related policy. This ensures that the voices of all stakeholders, especially those most affected by climate change, including marginalized and vulnerable communities, are heard and considered. A key element here is ensuring representation from any groups that a policy is designed to help. This means accessible consultations, communicated through the channels that will actually reach the groups in question, and being open to changing the model of a traditional consultation or stakeholder engagement process in order to facilitate and encourage opinions from anyone who would like to participate. It could look like writing new stakeholder engagement guidelines that prioritize research and outreach prior to organizing any type of engagement or consultation activity for example. This again is a sector agnostic recommendation and can be implemented in the public or the private sector by tailoring the process as needed.

Acknowledgements We would like to thank every MSME and NGO that took the time to speak with us in order to complete this chapter. Your time and tireless efforts in improving T&T are much appreciated. We would also like to thank Ryan Assiu and Nneka Mentore for sharing their valuable insights, helping us to shape this chapter, and supporting us throughout the process.

References

S Blackburn M Pelling 2018 The political impacts of adaptation actions: Social contracts, a research agenda Wiley Interdisciplinary Reviews: Climate Change 9 6 e549https://doi.org/10.1002/wcc.549

V Braun V Clarke 2006 Using thematic analysis in psychology Qualitative Research in Psychology 3 2 77 101 https://doi.org/10.1191/1478088706qp063oa

Bryman, A. (2012). *Social research methods* (4th ed.). Oxford University Press.

Central Statistical Office. (2024). *Population Statistics*. https://cso.gov.tt/subjects/population-and-vital-statistics/population/

L Gahman G Thongs A Greenidge 2021 Disaster, debt, and "underdevelopment": The cunning of colonial-capitalism in the Caribbean Development https://doi.org/10.1057/s41301-021-00282-4

Government of the Republic of Trinidad and Tobago (GORTT). (2021a). *National adaptation plan for the Republic of Trinidad and Tobago*. Ministry of Planning and Development.

Government of the Republic of Trinidad and Tobago (GORTT). (2021b). *Third National Communication of the Republic of Trinidad and Tobago to the United Nations Framework Convention on Climate Change*. Ministry of Planning and Development.

Hampton, S., Blundel, R., Wahga, A., Fawcett, T., & Shaw, C. (2022). Transforming small and medium-sized enterprises to address the climate emergency: The case for values-based engagement. *Corporate Social Responsibility and Environmental Management, 29*(5). https://doi.org/10.1002/csr.2279

International Institute for Sustainable Development. (2019). Caribbean Leaders Discuss Debt for Climate Adaptation Swap, Caribbean Resilience Fund. https://sdg.iisd.org/news/caribbean-leaders-discussdebt-for-climate-adaptation-swap-caribbean-resilience-fund/

International Labor Organization. (2023). *Perspectives on Business Resilience in Trinidad and Tobago: A Research Report*. https://www.ilo.org/sites/default/files/wcmsp5/groups/public/@americas/@ro-lima/@sro-port_of_spain/documents/publication/wcms_883104.pdf

Mees, H. L. P., Uittenbroek, C. J., Hegger, D. L. T., & Driessen, P. P. J. (2019). From citizen participation to government participation: An exploration of the roles of local governments in community initiatives for climate change adaptation in the Netherlands. *Environmental Policy and Governance, 29*(3). https://doi.org/10.1002/eet.1847

Ministry of Youth Development and National Service (MYDNS). (2024). *Enterprise development*. https://www.mydns.gov.tt/enterprise-development/

Montmasson-Clair, G., Patel, M., Mudombi, S., Jattansingh, S., Granderson, A., & Leotaud, N. (2019). *All is not green: Climate change adaptation and small business resilience in low-and middle-income countries*. Global Commission on Adaptation.

O'Brien, K., Hayward, B., & Berkes, F. (2009). Rethinking social contracts: Building resilience in a changing climate. *Ecology and Society, 14*(2). https://doi.org/10.5751/es-03027-140212

Plummer, N., Wilson, M., Yaneva-Toraman, I., McKenzie, C., Mitchell, S., Northover, P., Crowley, C., Edwards, T., & Richards, A. (2022). Recipes for resilience: Engaging Caribbean youth in climate action and food heritage through stories and song. *Sustainability, 14*(14), Article 8717. https://doi.org/10.3390/su14148717

Simon, K., Diprose, G., & Thomas, A. C. (2019). Community-led initiatives for climate adaptation and mitigation. Kōtuitui: *New Zealand Journal of Social Sciences Online, 15*(1), 93–105. https://doi.org/10.1080/1177083x.2019.1652659

United Nations Civil Society Unit. (n.d.). *About us*. https://www.un.org/en/civil-society/page/about-us

United Nations Framework Convention on Climate Change. (2021). *Resilience vision and summary*. https://unfccc.int/sites/default/files/resource/Resilience_Vision%26Summary.pdf

United Nations. (2024a). *Fast Facts: Small Island Developing States*. https://www.un.org/sites/un2.un.org/files/sids_fact_sheet_may2024.pdf

United Nations. (2024b). *About us*. United Nations Civil Society Unit. https://www.un.org/en/civil-society/page/about-us

World Bank. (2024). *Trinidad and Tobago Overview*. https://www.worldbank.org/en/country/trinidadandtobago/overview#:~:text=The%20economy%20is%20largely%20based,financial%20center%20in%20the%20Caribbean

World Economic Forum. (2024). *Global risks report 2024*. https://www.weforum.org/publications/global-risks-report-2024/digest/

CHAPTER 13

Climate Models Using AI—A Case for the OECS

Letetia M. Addison, Kevan Rajaram, Ken Manohar, and Curtis Charles

Introduction

The United Nations Sustainable Development Goals (SDGs) represent a blueprint to achieve a better and more sustainable future for all by addressing various global challenges, including climate change (United Nations Sustainable Development, 2015). SDG 13 encourages countries to take urgent action to combat climate change and its impact. Disaster preparedness due to climate change is a critical component of climate resilience and risk management in Caribbean islands. Small Island Developing States (SIDS) have their own peculiar vulnerabilities and are

L. M. Addison (✉) · K. Rajaram · K. Manohar
The University of the West Indies, St. Augustine, Trinidad and Tobago
e-mail: Letetia.Addison@sta.uwi.edu

C. Charles
The University of the West Indies, Five Islands Campus, Five Islands, Antigua and Barbuda
e-mail: curtis.charles@uwi.edu

D. D. Joseph et al. (eds.), *Climate Change*,
https://doi.org/10.1007/978-3-031-83037-2_13

expected to be impacted by more frequent and intense natural disasters due to climate change (sdgs.un.org, n.d.) .

SIDS within the Caribbean, such as those in the Organization of Eastern Caribbean States (OECS), are vulnerable to natural disasters due to their geographic location. These include storms and hurricanes during the annual rainy season and random events such as earthquakes and volcanic eruptions. These often result in damage to infrastructure, negative downturn to the economy and even loss of life.

Climate resilience can be defined as the ability to prepare for, recover from and adapt to the impacts from adverse weather conditions such as natural disasters (Centre for Climate and Energy Solutions, 2019). Increasing climate resilience is essential to mitigate the impact of a natural disaster and achieving this requires greater preparedness by each Caribbean country. Typical approaches can include infrastructural and financial protection via insurance (Muñoz & Ötker, 2018).

However, Information and Communication Technology (ICT) related solutions using Artificial Intelligence (AI) can also increase climate resilience. It is argued that Caribbean countries should incorporate "forecast-based early action" approaches to dealing with natural disasters (ReliefWeb, 2019). As a result, there is a need to adopt updated AI-based measures, alongside traditional approaches, to provide impactful and timely assistance for better disaster response and planning.

In most cases, the necessary data may be available which creates opportunities to incorporate Machine Learning (ML) models in disaster risk management. These AI models are data-driven, with the potential to significantly enhance disaster preparedness and response mechanisms.

Literature Review

The Latin American and Caribbean region is distinguished by its cultural multiplicity, extensive natural diversity and varied weather (Aide et al., 2013). Moreso, Small Island Developing States (SIDS) are acclaimed as a group of islands known by their small size, remoteness, limited resource base and vulnerability to global environmental concerns. Caribbean SIDS are low-lying coastal communities in tropical areas. While facing similar issues as the rest of the SIDS, Caribbean SIDS confront extra tropical challenges and limits, such as rising temperatures, sea level rise, flooding, drought and storms (Al-Tahir et al., 2011).

As a result, climate change has made this region more vulnerable to the consequences of extreme weather events, which are becoming more frequent and devastating (Eckstein et al., 2021). The Intergovernmental Panel on Climate Change (IPCC) published evidence in 2021 indicating that because global temperatures averaged more than 1.1 °C above historical trends during 2010 efforts to mitigate global warming alone are unlikely to be sufficient to counteract the negative consequences of climate change. Governments and individuals all across the world must work together to develop adaptation solutions (IPCC, 2022).

As such, the IPCC is currently using various AI-based climate models to forecast weather patterns, changes and anomalies around the world (IPCC, 2022). AI helps determine the most reliable models with the most accurate predictions and improves the accuracy of climate change projections. These ideas can be leveraged by Caribbean countries to curate actionable climate information to inform planning and policies. Moreso, ML "can be a powerful tool in reducing greenhouse gas emissions and assisting society in adapting to a changing climate", in part because "it has the potential to improve public services, help gather data for decision-making, and guide plans for future development" (Rolnick et al., 2022, pp. 2–3).

For instance, Mittapalli et al. (2021) proposed, NatDisP, which is a prospective application that uses wireless technology to anticipate up to five natural disasters depending on environmental factors and scenarios. The model involves building and fine-tuning machine learning and deep learning algorithms across multiple modules to predict disasters like earthquakes, floods and droughts. After extensive data preprocessing and algorithm comparison, the best-performing models are deployed in a user-friendly application that utilises real-time data collected via Internet of Things (IoT) technologies. It also makes recommendations for other natural calamities such as tsunamis. Furthermore, NatDisP emphasises the importance of leveraging wireless technologies to address the aftereffects of natural catastrophes. Using the support vector machine technique, the model by Mittapalli et al. (2021) attained an accuracy of 92.1%. The findings of Mittapalli et al. (2021) also imply that using natural features and signals provided by nature to predict natural disasters is superior to using location and timing data. The methods also eliminate the possibility of human errors or biases.

Additionally, landslide mapping and the reaction of this model on landslide susceptibility and landslide inventory mapping were investigated by Mutlu et al. (2019) via a deep learning approach (recurrent neural network). Results revealed that with a correct network structure and enough training iterations, the model tended to recognise the exact places of landslides with a very low percentage of false-positive predictions. At this point, the resulting model transitioned from susceptibility mapping to landslide inventory mapping. In other words, the deep learning algorithm's high capacity allowed for the establishment of a transition zone in terms of the capacity and performance of the landslide susceptibility model and the landslide inventory mapping model. The empirical findings of this study revealed that high-capacity deep learning algorithms enable semi-automatic generation of landslide inventory maps.

Moreover, Alemany et al. (2019), proposed a recurrent neural network (RNN) deployed over a grid system to contain the nonlinearity and complexity involved in forecasting storm tracks, potentially enhancing accuracy when compared to existing hurricane track forecasting models. Their model utilised latitude, longitude, wind speed and pressure that were publicly supplied by the National Hurricane Centre (NHC) to anticipate the route of a hurricane at 6-hour intervals. Data provided by Unisys Weather data was divided into 85% of the total hurricanes for training, and 15% for testing. Hurricanes or tropical storms were fed into the RNN one at a time. The results revealed that the suggested strategy is competitive with methods currently used by the NHC and can predict the path of a hurricane for up to 120 hours.

Furthermore, Saravi et al. (2019) examined the use of AI, particularly ML in natural disasters, more specifically flooding, to improve flood resilience and preparedness by estimating the flood type from the weather forecast, location, days event lasted, begin/end location, begin/end latitude and longitude, injuries direct/indirect, death direct/indirect and property and crop damage. Historical flooding data (1994–2018) was utilised from the Federal Emergency Management Agency (FEMA), National Oceanic and Atmospheric Administration (NOAA) and National Climatic Data Centre (NCDC). Several models such as Random Forest (RF), Lazy, J48 tree, Artificial Neural Network (ANN), Naive Bayes (NB) and Logistic Regression (LR) were created. These models were evaluated and the model that demonstrated the highest performance and accuracy in estimating flood types based on the given data was selected.

Saravi et al. (2019) demonstrated that Random Forest (RF) is a highly effective technique for classification tasks. Building on their findings, the best-performing technique used in Orange software was RF (based on the 164 test cases) with an RMSE result of 2.06, MAE of 0.19 and correctly classified instance rate of 80.49%. Likewise, the best-performing technique used in the Weka software (based on 3478 test cases) was also RF, with an RMSE result of 0.27, MAE of 0.1345 and correctly classified instance rate of 79.83%. As such, the classification results can lead to better decision-making concerning preventative and preparedness measures preparedness, and so to improve flood resilience. This is because most emergency responders can be alerted about the flood type that is likely to happen in the area and from this, local plans can be developed to lessen the threat and impact of flooding.

Together with this, Sun et al. (2020) provided an overview of current applications of AI in disaster management during its four phases of mitigation, preparedness, response and recovery. Concerning disaster mitigation, support vector machine (SVM) (Goetz et al., 2015; Xu et al., 2012; Zhou et al., 2018), logistic regression (LR) (Goetz et al., 2015; Zhou et al., 2018), random forest (RF) (Goetz et al., 2015) and neural networks (Dou et al., 2015; Zhou et al., 2018) can be used to estimate landslide susceptibility. The aforementioned AI algorithms have also been used to map the vulnerability of forest fires (Sachdeva et al., 2018), predict fire size (Mitsopoulos & Mallinis, 2017) and forecast precipitation (Huang & Xiang, 2018).

Additionally, AI approaches were used to anticipate potential consequences and identify vulnerabilities such that fragility curves, which were traditionally built from statistical analyses of historical and simulation data and can now be estimated from the application of AI methods such as LR (Ghosh et al., 2013; Kameshwar & Padgett 2014; Mangalathu et al., 2018a, 2018b), neural networks (Lagaros & Fragiadakis, 2007; Liu & Zhang, 2018; Mangalathu et al., 2018a, 2018b; Mitropoulou & Papadrakakis, 2011) and SVM (Mahmoudi & Chouinard, 2016). Generalised regression models (Liu et al., 2008; Reed, 2008), RF (Nateghi et al., 2014; Cerrai et al., 2019; D'Amico et al., 2019), decision tree (DT) (Wanik et al., 2015) and Bayesian additive regression tree (BART) (Cerrai et al., 2019) can also predict infrastructure service disruptions based on historical data.

Regarding disaster preparedness, in Europe, the Urban Flood project produced an internet-based network for early flood alerts, with an AI component designed for detecting abnormal dike behaviour based on the analysis of thousands of sensor streams (Noymanee et al., 2017). Sakaki et al. (2013) even performed semantic analysis of Japanese tweets using a tweet crawler, predicted the earthquake location and produced Toretter, a reporting system that was faster than Japan Meteorological Agency broadcasts.

Utility firms can employ AI-based technologies to anticipate likely damaged areas and service interruption durations prior to a disaster event and prepare accordingly. For example, Hydro One, a prominent utility company in Ontario, Canada, successfully employed such real-time predictive assessments to place employees in important areas and effectively restore electricity supply within four days, considerably shortening the restoration time (McConnon, 2018). Thus, SVM, DT, neural networks, reinforcement learning and optimisation algorithms are among the most commonly used AI approaches for evacuations (Sun, Bocchini & Davison, 2020).

In terms of disaster response, maps of infrastructure inventory models (Eguchi et al. 2008), damaged buildings and bridges (Adams et al., 2002; Balz & Liao, 2010; Hutchinson & Chen, 2005) and disaster-impacted regions (Casagli et al., 2017; Rosser et al., 2017) have been created using satellite photos. Classifier algorithms are frequently employed for analysing these maps and photos (Vetrivel et al., 2016). Moreover, machine listening can aid in the automatic recognition of voices, the identification of keywords with high priority and the quick processing of speech data from various places (Ramchurn et al., 2016).

Sentiment mining, using natural language processing algorithms, can assist disaster managers in crisis management and enable effective disaster relief by providing greater understanding of the situation, such as where to dispatch first responders and distribute resources. There are also numerous applications for disaster information systems that extract information from social media data (Neppalli et al., 2018), mobile phone data (Sun & Tan, 2019), remote sensing data and aerial photos (Morito et al., 2016; Tian & Chen, 2017) utilising supervised models and deep learning.

Relating to disaster recovery, by using aerial photos, social media imagery data and sensor measurement data, AI algorithms can help eliminate human efforts in visual inspection (Khaloo et al., 2017; Khoshnoudian et al., 2017). Furthermore, sentiment analysis of social media data can track human activity patterns throughout the recovery process when measuring the disaster's influence on humans (Caragea et al., 2014; Hasan & Ukkusuri, 2014; Liu et al., 2020; Resch et al., 2018; Shelton et al., 2014).

With surveys being the main method when studying psychological distress following a tragedy, these results are frequently analysed using both supervised and unsupervised models, particularly regression methods, dimension reduction methods and neural networks to identify risk factors and assess the effectiveness of preventive interventions (Gao et al., 2006; Gong et al., 2014; Huang et al., 2010; Kim et al., 2008; Rosellini et al., 2018).

AI approaches can aid in the estimation of disaster losses and rehabilitation costs. Specifically, supervised models, in particular, regression and neural networks, have been used to rapidly process imagery for detecting structural damage, identifying repair needs and estimating repair costs; they have also been used to analyse historical dispersion data of disaster recovery funds for budget allocations, and process insurance claims in less time (Barthel & Neumayer, 2012; Chen & Huang, 2006; Stojadinovic et al., 2017; Zagorecki et al., 2013).

However, AI in the Caribbean is an under-utilised technology, especially in climate resilience. Existing climate resilience efforts are varied, including social and economic strategies implemented at local and national levels. AI models can greatly enhance climate resilience, even at regional levels. The main objective is to improve climate resilience in the Caribbean by providing an AI-based tool for Disaster and Emergency Response teams to predict the likelihood of natural disasters occurring. This uses ML Classification techniques, particularly in the OECS. Using various databases, classification models are built with country-level variables based on climate data (including *Temperature, Precipitation, Historical Natural disasters*).

The key objectives of this tool are:

- To **build a robust classification model** on country-level data to predict risks of natural disasters in various regions.
- To **design a scalable tool** to provide risk predictions for natural disasters in OECS countries.
- To **promote the tool as a User Interface** which related government, climate and disaster risk agencies and the public can access for predictive insights ahead of natural disaster events.

On a larger scale, the predictions of natural disaster occurring in a specific region can be used to:

- **Improve emergency response contingencies** ahead of time for more efficient optimisation of aid.
- **Provide a planning tool for policymakers** in the Regional Consortium such as in the areas of agriculture, tourism and other key stakeholders.
- **Educate the public** about most vulnerable regions for improved personal and property protection measures.

Methodology

Predictive Analytics is the application of machine learning algorithms and statistics on large data sets to predict or estimate future outcomes (Larose & Larose, 2015). It involves analyses of data using statistical modelling techniques. This approach was employed in order to extract the relevant metrics for the model prototype.

OECS Disaster Risk Smart Classifier (DRSmC)

DRSmC is a multifaceted tool which uses an AI approach using ML classification models to predict the likelihood of a natural disaster occurring in the OECS. It leverages historical climate data to calculate a probabilistic risk value corresponding to occurrence of the natural disaster event, such as a flood. This value is categorised into a risk band or likelihood of low, medium, high or very high.

The Base Model: Proof of Concept (POC)

A Proof-of-Concept (POC) of the tool was created to demonstrate its application to AI-based climate resilience in the Caribbean. The scope was narrowed to predicting the likelihood of a natural disaster event occurring in selected OECS countries based on historical data for temperature and rainfall. The natural disaster events of interest were flood, storm or drought. The countries chosen for this study were Antigua and Barbuda, Dominica, Grenada, Montserrat, St. Kitts and Nevis, St. Lucia and St. Vincent and the Grenadines, due to the availability of sufficient data from these locations.

Data Preparation

The data was sourced from the World Bank Group (World Bank, 2022) and other public datasets. The data was cleaned and prepared using Microsoft's Power Query Editor and then collated into one master file with data ranging from 1901 to 2020, monthly. In the base model, each country has independent variables related to climate including Year, Month, Date, Average Temperature, Minimum Temperature, Maximum Temperature and Rainfall (Precipitation). A binary target and dependent variable were created to represent the occurrence or non-occurrence of a natural disaster event.

Model Building

Model training, testing and validation were performed using thirteen (13) different ML algorithms (Random Forest, Gradient Boosted Trees, Logistic Regression, LightGBM, XGBoost, Decision Tree, K Nearest Neighbours (k = 5), Extra trees, Artificial Neural Network, SGD, LASSO-LARS, SVM). The train/test split was 80/20. The best-performing model for each country is shown in Table 13.1, with technical aspects including various model metrics to support the accuracy.

Table 13.1 Model metrics for country-level data in OECS countries

	Antigua	*Dominica*	*Grenada*	*Montserrat*	*St. Kitts & Nevis*	*St. Lucia*	*St. Vincent & Grenadines*
Best model	Logistic regression	Extra trees	LASSO-LARS	Logistic regression	Artificial neural network	K-Nearest neighbours	Support vector machine
F1-Score	0.5	0.22	0.33	0.571	0.6	0.55	0.5
Precision	0.5	0.17	0.2	1.0	0.6	0.43	0.8
Recall	0.5	0.33	1.0	0.4	0.6	0.75	0.36
Accuracy	99%	97%	99%	100%	99%	98%	99%

User Interface

The results from these models were visualised using Power BI to showcase the main contributors to natural disaster and the risk of the occurrence of the natural disaster event. Figure 13.1 shows the DRSmC User Interface prototype which provides information on the key influencers based on specific country-level climate variables, such as temperature and rainfall by month.

The risk levels associated with each of the variable ranges for temperature and rainfall as well as certain months and seasons for each country are also shown in a bar chart for more granular insights. For instance, there is an 80% chance of a flood event for an average temperature range of 22.3–24.5 °C in Antigua and Barbuda according to the model. This is regarded as "high risk" as the chance of a flood increases in this range. As such, these predictive aspects are key to providing forecast-based early action for the islands in the OECS individually and overall. Moreover, this predictive model, which assesses flood risk based on temperature and rainfall ranges, can be extended for use in other Caribbean countries due to its versatility and robustness.

Such insights for the OECS are provided in the POC prototype report, outlining the winning solution for the Growth & Resilience Dialogue (GRD) 2022 Climate Resilience Data Challenge, involving the development of AI Application Prototypes on key Climate Data Issues in the OECS region. This is published online at: https://sites.google.com/view/grd-climate-challenge-2022/proof-of-concept and is available for demonstration upon request.

Predicting Likelihood of a Natural Disaster

For easier consumption by Disaster and Emergency Response teams, as well as the public, the probability risk of a natural disaster event was converted into a likelihood of occurrence. This conversion can be seen in Table 13.2:

These risk bands provide an intuitive sense of the level of preparation needed for a pending natural disaster event.

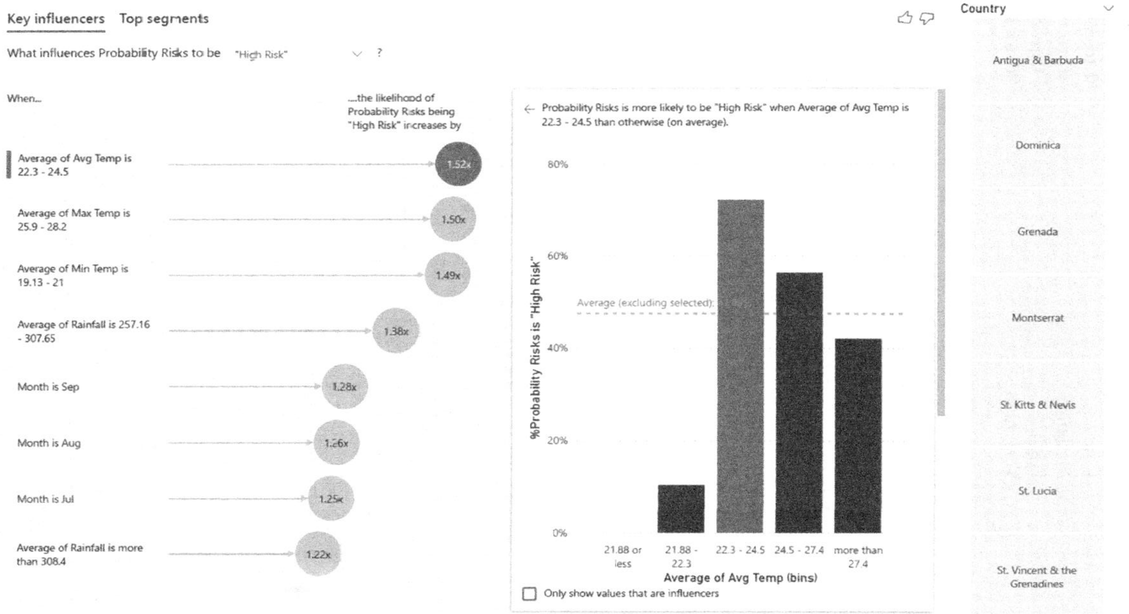

Fig. 13.1 Predictive model visualised in Microsoft Power BI

Table 13.2 Natural Disaster Risk Bands

Likelihood of occurrence	*Probability risk*
Low	0 to 39%
Medium	40 to 69%
High	70 to 89%
Very high	90 to 100%

Expected Outcomes and Considerations

Scalability of the tool is important since the base model uses data from OECS regions which can be extended to other regions in the Caribbean. Data availability may vary in different countries and this tool provides a base platform for experimentation and model validation for suitability in other regions.

Some further considerations for this work include:

- **Policy adoption by various agencies** through benefit-based optimisation models to convince stakeholders to adapt to new actions and policies based on results from AI methods (forecasting, classification, optimisation).
- **Scientific research and publication** with datasets being made publicly available with relevant permissions where possible.
- **Capacity-building activities** such as community outreach, engagement activities such as conferences, webinars to sensitise and communicate with stakeholders about relevant aspects of this project.
- **Creation of a singular digital platform** to host the various datasets obtained through this project and that can act as a database to inform on further research and analyses is also important.

Conclusions and Future Insights

The OECS DRSmC tool is a data-driven advisory mechanism utilising AI modelling to accurately predict natural disaster occurrences. By analysing historical data, the tool classifies risk levels for impending natural disasters, offering valuable insights for disaster response agencies to mitigate potential losses. The base model's predictive capabilities can be enhanced with additional country-level historical, spatiotemporal, economic and social media data.

Future research could greatly enhance the OECS DRSmC tool's effectiveness by integrating additional data sources such as land-use patterns, and satellite imagery, which would improve risk assessments. Expanding the tool's spatiotemporal analysis with Geographic Information Systems (GIS) and remote sensing data could refine predictions by accounting for how risks evolve over time and across regions. Advanced machine learning techniques, including deep learning and hybrid models, could reveal complex data patterns and improve prediction accuracy, while real-time data integration would provide critical, up-to-the-minute risk assessments. Customising the tool for various stakeholders and conducting longitudinal studies could assess its impact on disaster preparedness and response. Additionally, adapting the model for other regions and incorporating climate change scenarios could enhance its applicability and support long-term planning, contributing to more resilient and prepared communities.

Some future applications within the Regional Consortium may include:

- **Agricultural Warning Systems**: Farmers and other related stakeholders can be advised ahead of time to rearrange planting schedules to reduce economic losses.
- **Tourism and Construction Sector Advisory Systems**: Hotels, guest houses and other construction stakeholders can have a schedule ahead of time about specific areas and times of year which may have higher predicted risks of natural disaster impacts and thus reduce economic losses.
- **General Emergency and Disaster Response Systems**: These agencies can use this information to improve prioritisation and timely coordination of relief efforts in the most affected areas.

On a broader scale, the model risk predictions can optimise the distribution of relief materials, assisting in the timely support of flood victims and reducing physical and economic losses. While the primary focus is on aiding emergency responders, the tool also has the potential to alert policymakers and the public to the increasing frequency of climate-related disasters, driving the need for immediate and proactive measures. Given the urgent timeframe for corrective action (United Nations, 2019), showcasing localised impacts could motivate policymakers and invigorate the public to demand more effective climate resilience strategies. Thus, the OECS DRSmC tool can significantly improve disaster risk prediction and response, thereby enhancing resilience and preparedness in vulnerable communities while providing crucial insights for long-term climate adaptation and policy development.

References

Adams, B. J., Huyck, C., Mansouri, B., Eguchi, R., & Shinozuka, M. (2002). Post-disaster bridge damage assessment [Conference presentation]. In *Proceedings of the 15th Pecora Conference: Integrating Remote Sensing at the Global, Regional, and Local Scale*, Denver, CO, United States.

M Aide M Clark R Grau D López-Carr M Levy D Redo M Bonilla-Moheno G Riner M Andrade-Núñez M Muñiz 2013 Deforestation and reforestation of Latin America and the Caribbean (2001–2010) Biotropica 45 2 262 271 https://doi.org/10.1111/j.1744-7429.2012.00908.x

S Alemany J Beltran A Perez S Ganzfried 2019 Predicting hurricane trajectories using a recurrent neural network Proceedings of the AAAI Conference on Artificial Intelligence 33 1 468 475 https://doi.org/10.1609/aaai.v33i01.3301468

Al-Tahir, R., Arthur, M., & Davis, D. (2011). Low-cost aerial mapping alternatives for natural disasters in the Caribbean (# 5153) [Conference presentation]. In *FIG Working Week 2011 Bridging the Gap between Cultures*, Marrakech, Morocco.

T Balz M Liao 2010 Building-damage detection using post-seismic high-resolution SAR satellite data International Journal of Remote Sensing 31 13 3369 3391 https://doi.org/10.1080/01431161003727671

F Barthel E Neumayer 2012 A trend analysis of normalised insured damage from natural disasters Climatic Change 113 2 215 237 https://doi.org/10.1007/s10584-011-0331-2

Casagli, N., Frodella, W., Morelli, S., Tofani, V., Ciampalini, A., Intrieri, E., Raspini, F., Rossi, G., Tanteri, L., & Lu, P. (2017). Spaceborne, UAV and ground-based remote sensing techniques for landslide mapping, monitoring and early warning. *Geoenvironmental Disasters*, *4*(1), Article 9. https://doi.org/10.1186/s40677-017-0073-1

Caragea, C., Squicciarini, A., Stehle, S., Neppalli, K., & Tapia, A. (2014). Mapping moods: Geo-mapped sentiment analysis during Hurricane Sandy

[Conference presentation]. In *Proceedings of the 11th International Conference on Information Systems for Crisis Response and Management*, University Park, PA, United States.

Center for Climate and Energy Solutions. (2019). *What is climate resilience, and why does it matter?* https://www.c2es.org/document/what-is-climate-resilience-and-why-does-it-matter/

D Cerrai DW Wanik MAE Bhuiyan X Zhang J Yang MEB Frediani EN Anagnostou 2019 Predicting storm outages through new representations of weather and vegetation IEEE Access 7 29639 29654 https://doi.org/10.1109/ACCESS.2019.2902558

WT Chen YH Huang 2006 Approximately predicting the cost and duration of school reconstruction projects in Taiwan Construction Management and Economics 24 12 1231 1239 https://doi.org/10.1080/01446190600953805

DF D'Amico SM Quiring CM Maderia DB McRoberts 2019 Improving the hurricane outage prediction model by including tree species Climate Risk Management 25 100193https://doi.org/10.1016/j.crm.2019.100193

J Dou H Yamagishi HR Pourghasemi AP Yunus X Song Y Xu Z Zhu 2015 An integrated artificial neural network model for the landslide susceptibility assessment of Osado Island Japan. Natural Hazards 78 3 1749 1776 https://doi.org/10.1007/s11069-015-1799-2

Eckstein, D., Künzel, V., & Schäfer, L. (2021). Global climate risk index 2021. *Who suffers most from extreme weather events, 2000–2019*. Germanwatch. www.germanwatch.org/en/cri

Eguchi, R. T., Huyck, C. K., Ghosh, S., & Adams, B. J. (2008). The application of remote sensing technologies for disaster management. In *Paper Presented at the 14th World Conference on Earthquake Engineering*, Paper ID K004. https://www.iitk.ac.in/nicee/wcee/article/14_K004.pdf

Y Gao YX Chen YS Ding BY Tang 2006 Immune gentic algorithm based on network model for flood disaster evaluation Journal of Natural Disaster 15 110 114

J Ghosh JE Padgett LD Dueñas-Osorio 2013 Surrogate modelling and failure surface visualisation for efficient seismic vulnerability assessment of highway bridges Probabilistic Engineering Mechanics 34 189 199 https://doi.org/10.1016/j.probengmech.2013.09.003

JN Goetz A Brenning H Petschko P Leopold 2015 Evaluating machine learning and statistical prediction techniques for landslide susceptibility modelling Computers & Geosciences 81 1 11 https://doi.org/10.1016/j.cageo.2015.04.007

Q Gong L Li S Tognin Q Wu W Pettersson-Yeo S Lui X Huang AF Marquand A Mechelli 2014 Using structural neuroanatomy to identify trauma survivors

with and without post-traumatic stress disorder at the individual level Psychological Medicine 44 1 195 203 https://doi.org/10.1017/S0033291713000561
S Hasan SV Ukkusuri 2014 Urban activity pattern classification using topic models from online geo-location data Transportation Research Part C: Emerging Technologies 44 363 381 https://doi.org/10.1016/j.trc.2014.04.003
L Huang L Xiang 2018 Method for meteorological early warning of precipitation-induced landslides based on deep neural network Neural Processing Letters 48 2 1243 1260 https://doi.org/10.1007/s11063-017-9778-0
Z Huang J Zhou L Song Y Lu Y Zhang 2010 Flood disaster loss comprehensive evaluation model based on optimization support vector machine Expert Systems with Applications 37 5 3810 3814 https://doi.org/10.1016/j.eswa.2009.11.039
TC Hutchinson Z Chen 2005 Optimised estimated ground truth for object-based urban damage estimation using satellite images from the 2003 Bam, Iran, Earthquake Earthquake Spectra 21 239 254 https://doi.org/10.1193/1.2098937
Intergovernmental Panel on Climate Change (IPCC). (2021). *Summary for policymakers*. In V. Masson-Delmotte, P. Zhai, A. Pirani, S. L. Connors, C. Péan, S. Berger, N. Caud, Y. Chen, L. Goldfarb, M. I. Gomis, M. Huang, K. Leitzell, E. Lonnoy, J. B. R. Matthews, T. K. Maycock, T. Waterfield, O. Yelekçi, R. Yu, & B. Zhou (Eds.), *Climate change 2021: The physical science basis. Contribution of Working Group I to the Sixth Assessment Report of the Intergovernmental Panel on Climate Change*. https://www.ipcc.ch/report/ar6/wg1/downloads/report/IPCC_AR6_WGI_SPM_final.pdf
Intergovernmental Panel on Climate Change (IPCC). (2022). *Summary for policymakers*. In P. R. Shukla, J. Skea, R. Slade, A. Al Khourdajie, R. van Diemen, D. McCollum, M. Pathak, S. Some, P. Vyas, R. Fradera, M. Belkacemi, A. Hasija, G. Lisboa, S. Luz, & J. Malley (Eds.), *Climate change 2022: Mitigation of climate change. Contribution of Working Group III to the Sixth Assessment Report of the Intergovernmental Panel on Climate Change*. Cambridge University Press. https://www.ipcc.ch/report/ar6/wg3/downloads/report/IPCC_AR6_WGIII_SPM.pdf
Kameshwar, S., & Padgett, J. E. (2014). Multi-hazard risk assessment of highway bridges subjected to earthquake and hurricane hazards. *Engineering Structures*, *78*, 154–166. https://doi.org/10.1016/j.engstruct.2014.05.016
A Khaloo D Lattanzi K Cunningham R Dell'Andrea M Riley 2017 Unmanned aerial vehicle inspection of the Placer River trail bridge through image-based 3D modelling Structure and Infrastructure Engineering 14 1 124 136 https://doi.org/10.1080/15732479.2017.1330891

Khoshnoudian, F., Talaei, S., & Fallahian, M. (2017). Structural damage detection using FRF data, 2D-PCA, artificial neural networks and imperialist competitive algorithm simultaneously. *International Journal of Structural Stability and Dynamics*, *17*(7). https://doi.org/10.1142/S0219455417500730

D Kim H-J Lee S Cho 2008 Response modelling with support vector regression Expert Systems with Applications 34 2 1102 1108 https://doi.org/10.1016/j.eswa.2006.12.019

ND Lagaros M Fragiadakis 2007 Fragility assessment of steel frames using neural networks Earthquake Spectra 23 4 735 752 https://doi.org/10.1193/1.2798241

DT Larose CD Larose 2015 Data mining and predictive analytics John Wiley & Sons

H Liu RA Davidson TV Apanasovich 2008 Spatial generalised linear mixed models of electric power outages due to hurricanes and ice storms Reliability Engineering & System Safety 93 6 897 912 https://doi.org/10.1016/j.ress.2007.03.038

Z Liu Y Du J Yi F Liang T Ma T Pei 2020 Quantitative estimates of collective geo-tagged human activities in response to typhoon Hato using location-aware big data International Journal of Digital Earth 13 9 1072 1092 https://doi.org/10.1080/17538947.2019.1645894

Z Liu Z Zhang 2018 Artificial neural network-based method for seismic fragility analysis of steel frames KSCE Journal of Civil Engineering 22 2 708 717 https://doi.org/10.1007/s12205-017-1329-8

SN Mahmoudi L Chouinard 2016 Seismic fragility assessment of highway bridges using support vector machines Bulletin of Earthquake Engineering 14 6 1571 1587 https://doi.org/10.1007/s10518-016-9894-7

Mangalathu, S., Jeon, J. S., & DesRoches, R. (2018a). Critical uncertainty parameters influencing seismic performance of bridges using Lasso regression. *Earthquake Engineering & Structural Dynamics*, *47*(3), 784–801. https://doi.org/10.1002/eqe.2991

Mangalathu, S., Heo, G., & Jeon, J. S. (2018b). Artificial neural network-based multi-dimensional fragility development of skewed concrete bridge classes. *Engineering Structures*, *162*, 166–176. https://doi.org/10.1016/j.engstruct.2018.01.053

McConnon, A. (2018). AI helps cities predict natural disasters. *Wall Street Journal*. https://www.wsj.com/articles/ai-helps-cities-predict-natural-disasters-1530065100

I Mitsopoulos G Mallinis 2017 A data-driven approach to assess large fire size generation in Greece Natural Hazards 88 3 1591 1607 https://doi.org/10.1007/s11069-017-2934-z

CC Mitropoulou M Papadrakakis 2011 Developing fragility curves based on neural network IDA predictions Engineering Structures 33 12 3409 3421 https://doi.org/10.1016/j.engstruct.2011.07.005

Mittapalli, J. S., Mutha, J. A., & R, M. (2021). *NatDisP—An intelligent natural disaster predictor*. Research Square. https://doi.org/10.21203/rs.3.rs-204305/v1

Morito, T., Sugiyama, O., Kojima, R., & Nakadai, K. (2016). Partially shared deep neural network in sound source separation and identification using a UAV-embedded microphone array. In *IEEE/RSJ International Conference on Intelligent Robots and Systems (IROS)* [Conference presentation], (pp. 1299–1304). https://doi.org/10.1109/IROS.2016.7759215

Muñoz, S., & Ötker, İ. (2018). Building resilience to natural disasters in the Caribbean requires greater preparedness. *International Monetary Fund*. https://www.imf.org/en/News/Articles/2018/12/07/NA120718-Building-Resilience-to-Natural-Disasters-in-Caribbean-Requires-Greater-Preparedness

B Mutlu HA Nefeslioglu EA Sezer MA Akcayol C Gokceoglu 2019 An experimental research on the use of recurrent neural networks in landslide susceptibility mapping ISPRS International Journal of Geo-Information 8 12 578 https://doi.org/10.3390/ijgi8120578

R Nateghi SD Guikema SM Quiring 2014 Power outage estimation for tropical cyclones: Improved accuracy with simpler models Risk Analysis 34 6 1069 1078 https://doi.org/10.1111/risa.12131

Neppalli, V. K., Caragea, C., & Caragea, D. (2018). Deep neural networks versus Naive Bayes classifiers for identifying informative tweets during disasters [Conference presentation]. In K. Boersma & B. Tomaszewski (Eds.), *Proceedings of the 15th International Conference on Information Systems for Crisis Response and Management*. Association for Computing Machinery. http://idl.iscram.org/

J Noymanee NO Nikitin AV Kalyuzhnaya 2017 Urban pluvial flood forecasting using open data with machine learning techniques in Pattani Basin Procedia Computer Science 119 288 297 https://doi.org/10.1016/j.procs.2017.11.187

SD Ramchurn TD Huynh F Wu Y Ikuno J Flann L Moreau JE Fischer W Jiang T Rodden E Simpson S Reece S Roberts NR Jennings 2016 A disaster response system based on human-agent collectives Journal of Artificial Intelligence Research 57 661 708 https://doi.org/10.1613/jair.5098

D Reed 2008 Electric utility distribution analysis for extreme winds Journal of Wind Engineering & Industrial Aerodynamics 96 1 123 140 https://doi.org/10.1016/j.jweia.2007.04.002

ReliefWeb. (2019). *OECS partners convene in Saint Lucia to assess how to improve disaster preparedness*. *ReliefWeb*. https://reliefweb.int/report/saint-lucia/oecs-partners-convene-saint-lucia-assess-how-improve-disaster-preparedness

D Rolnick PL Donti HL Kaack K Kochanski A Lacoste K Sankaran AS Ross N Milojevic-Dupont N Jaques A Waldman-Brown AS Luccioni T Maharaj ED Sherwin SK Mukkavilli KP Kording CP Gomes AY Ng D Hassabis JC Platt F Creutzig J Chayes Y Bengio 2022 Tackling climate change with machine learning ACM Computing Surveys 55 2 1 96 https://doi.org/10.1145/3485128

B Resch F Usländer C Havas 2018 Combining machine-learning topic models and spatiotemporal analysis of social media data for disaster footprint and damage assessment Journal of Artificial Intelligence Research 45 4 362 376 https://doi.org/10.1080/15230406.2017.1356242

AJ Rosellini F Dussailant JR Zubizarreta RC Kessler S Rose 2018 Predicting post-traumatic stress disorder following a natural disaster Journal of Psychiatric Research 96 15 22 https://doi.org/10.1016/j.jpsychires.2017.09.010

JF Rosser DG Leibovici MJ Jackson 2017 Rapid flood inundation mapping using social media, remote sensing and topographic data Natural Hazards 87 103 120 https://doi.org/10.1007/s11069-017-2755-0

T Sakaki M Okazaki Y Matsuo 2013 Tweet analysis for real-time event detection and earthquake reporting system development IEEE Transactions on Knowledge and Data Engineering 25 4 919 931 https://doi.org/10.1109/TKDE.2012.29

Saravi, S., Kalawsky, R., Joannou, D., Casado, M. R., Fu, G., & Meng, F. (2019). Use of artificial intelligence to improve resilience and preparedness against adverse flood events. *Water*, *11*(5), Article 973. https://doi.org/10.3390/w11050973

S Sachdeva T Bhatia AK Verma 2018 GIS-based evolutionary optimised gradient boosted decision trees for forest fire susceptibility mapping Natural Hazards 92 3 1399 1418 https://doi.org/10.1007/s11069-018-3256-5

T Shelton A Poorthuis M Graham M Zook 2014 Mapping the data shadows of Hurricane Sandy: Uncovering the sociospatial dimensions of big data Geoforum 52 167 179 https://doi.org/10.1016/j.geoforum.2014.01.006

Stojadinovic, Z., Kovacevic, M., Marinkovic, D., & Stojadinovic, B. (2017). Data-driven housing damage and repair cost prediction framework based on the 2010 Kraljevo earthquake data [Paper presentation]. In *Proceedings of the 16th World Conference on Earthquake Engineering (16WCEE)*, 4987. https://www.wcee.nicee.org/wcee/article/16WCEE/WCEE2017-4987.pdf

Sun, W., Bocchini, P., & Davison, B. D. (2020). Applications of artificial intelligence for disaster management. *Natural Hazards: Journal of the International Society for the Prevention and Mitigation of Natural Hazards*, *103*(3). Springer; International Society for the Prevention and Mitigation of

Natural Hazards, (pp. 2631–2689). https://doi.org/10.1007/s11069-020-04124-3

Sun, Y., & Tan, W. (2019). A trust-aware task allocation method using deep Q-learning for uncertain mobile crowdsourcing. *Human-centric Computing and Information Sciences*, *9*(1). https://doi.org/10.1186/s13673-019-0187-4

sdgs.un.org. (n.d.). *Small Island Developing States | Department of Economic and Social Affairs*. https://sdgs.un.org/topics/small-island-developing-states

Tian, H., & Chen, S. C. (2017). A video-aided semantic analytics system for disaster information integration [Conference presentation]. In *The 3rd IEEE International Conference on Multimedia Big Data*, (pp. 242–243). IEEE. https://doi.org/10.1109/BigMM.2017.31

United Nations Sustainable Development. (2015). *About the sustainable development goals*. https://www.un.org/sustainabledevelopment/sustainable-development-goals/

United Nations. (2019). *Only 11 years left to prevent irreversible damage from climate change, speakers warn during general assembly high-level meeting*. Meetings Coverage and Press Releases. https://www.un.org/press/en/2019/ga12131.doc.htm

Vetrivel, A., Kerle, N., Gerke, M., Nex, G., & Vosselman, G. (2016). Towards automated satellite image segmentation and classification for assessing disaster damage using data-specific features with incremental learning. In N. Kerle, M. Gerke, & S. Lefevre (Eds.), *Proceedings of the GEOBIA2016: Solutions and synergies* (pp. 1–5). University of Twente, Faculty of Geo-Information Science and Earth Observation (ITC). https://doi.org/10.3990/2.369

DW Wanik EN Anagnostou BM Hartman MEB Frediani M Astitha 2015 Storm outage modelling for an electric distribution network in Northeastern USA Natural Hazards 79 2 1359 1384 https://doi.org/10.1007/s11069-015-1908-2

World Bank. (2022). *World bank open data | Data*. https://data.worldbank.org/

C Xu F Dai X Xu Y-H Lee 2012 GIS-based support vector machine modelling of earthquake-triggered landslide susceptibility in the Jianjiang River watershed, China Geomorphology 145–146 70 80 https://doi.org/10.1016/j.geomorph.2011.12.040

AT Zagorecki DE Johnson J Ristvej 2013 Data mining and machine learning in the context of disaster and crisis management International Journal of Emergency Management 9 4 351 365 https://doi.org/10.1504/IJEM.2013.059879

C Zhou K Yin Y Cao B Ahmed Y Li F Catani HR Pourghasemi 2018 Landslide susceptibility modelling applying machine learning methods: A case study from Longju in the Three Gorges reservoir area, China Computers & Geosciences 112 23 37 https://doi.org/10.1016/j.cageo.2017.11.019

CHAPTER 14

Conclusion

Camille Huggins, Debra D. Joseph, and Maud Mthembu

There is one thing that is undeniable, climate change is causing irreparable damage to the world. The Global South is disproportionately vulnerable to the impact of climate change despite contributing the least to global emissions. Nations in the Global South face heightened vulnerability to extreme weather events like floods, droughts, and hurricanes due to poor infrastructure, high population densities, and limited resources. These constraints hinder their ability to implement effective mitigation and adaptation strategies. Additionally, their economies are frequently susceptible to climate shocks as most of their industries are climate-sensitive, such as agriculture, tourism, and the harvesting of natural resources.

C. Huggins (✉)
The University of the West Indies, St. Augustine, Trinidad and Tobago
e-mail: camille.huggins@uwi.edu

D. D. Joseph
The University of the West Indies, Cave Hill, Bridgetown, Barbados
e-mail: debra.joseph@cavehill.uwi.edu

M. Mthembu
University of KwaZulu-Natal, Pietermaritzburg, South Africa
e-mail: mthembum4@ukzn.ac.za

D. D. Joseph et al. (eds.), *Climate Change*,
https://doi.org/10.1007/978-3-031-83037-2_14

The impact of climate change on industries fueling the economy such as agriculture is projected to decline by 10–25%, resulting in billions of dollars in lost revenue. The tourism industry is also under threat from coastal erosion and reef damage, which undermines the natural attractions that sustain these economies. The financial burden of climate change on Global South nations includes the direct economic losses as well as the preparation to guard against climate events. The infrastructure damage due to extreme weather events such as hurricanes, floods, and droughts can be as much as $290 and upwards to $580 billion by 2030, and this can rise to 1.8 trillion by 2050.

Climate change poses a significant barrier to the ongoing development of many developing countries, severely hindering progress toward achieving and sustaining the Sustainable Development Goals (SDGs). Goals such as poverty reduction, hunger eradication, and ensuring access to quality education are particularly at risk. The devastating impact of climate change inflicts profound psychological and social trauma across Global South communities, creating mental health challenges and social disruption.

The mental health burden of climate change is profound and multifaceted, affecting individuals and communities in diverse ways. Many people experience acute trauma from extreme weather events such as hurricanes or floods, while others endure eco anxiety about future climate threats and their implications. The loss of livelihoods, particularly in climate-sensitive sectors like agriculture or tourism, often leads to depression and feelings of hopelessness. Additionally, intergenerational trauma emerges as traditional knowledge about the environment becomes less reliable, eroding cultural practices and connections. Traditional communities, whose identities and cultural practices are deeply intertwined with their physical environments, face not just physical displacement but also a devastating loss of cultural continuity and ancestral connections. Among the youth, climate anxiety is increasingly prevalent, as they grapple with the uncertainty of their future in a rapidly changing world.

Climate change significantly disrupts communities, leading to profound social and cultural consequences. In resource deprived areas, there is an alarming rise in gender-based violence, as scarcity heightens tensions and exacerbates existing inequalities. Forced migration due to displacement also contributes to the erosion of community cohesion, as people are uprooted from their homes and forced to rebuild their lives in unfamiliar settings. Moreover, conflicts over diminishing resources such as water, arable land, and food have become increasingly common, further straining relationships within and between communities.

This timely edited volume highlights the pressing challenges of climate change and offers practical interventions and strategies for mitigating the effects of climate change in the Global South. Over the past few decades, climate change has become a critical concern for countries in the Global South, who are vulnerable and unprepared to address present and future consequences.

As an edited volume, *The Impacts of Climate Change* explores the unique needs of vulnerable populations and proposes actionable solutions for creating resilient and safe spaces in the face of environmental crises. Beyond identifying key issues for sustainable development, the book underscores the importance of resilience in navigating climate challenges and offered practical insights for scholars and practitioners working on the ground. The subsequent chapters delve into the impact of climate change across various regions, including South Africa, Barbados, Trinidad and Tobago, South America, and South Asia. These chapters shed light on the psychosocial effects of climate change, exploring critical issues such as migration, food security, natural disasters, and their profound implications for affected populations. By examining these regions, the book underscores the diverse, yet interconnected challenges faced by communities as they navigate the complex realities of a changing climate. The book is structured into three main sections, each addressing key topics with depth and clarity. These sections are designed to guide readers through the most salient and pertinent themes, providing a cohesive and comprehensive exploration of the subject matter.

The first section lays a foundational framework, offering critical context on the multifaceted impact of climate change in both South Africa and the Caribbean. It delves into the intricate interplay between environmental challenges and human psychology. This section explores how climate change not only reshapes ecosystems and economies but also profoundly affects mental and emotional well-being. This section sets the stage for a nuanced understanding of the global and regional implications of climate-induced transformations.

The chapter, *Why Climate Change Matters for Vulnerable Populations in South Africa: A Social Work Perspective* written by Nel and Khosa, explores the intersection of climate change and social work within South Africa, focusing on its impact on vulnerable populations in particular women, children and those living in poverty. Climate change exacerbates existing socio-economic inequalities, intensifying the challenges. Women, often bearing the brunt of systemic oppression and reduced access to resources, are disproportionately affected by environmental degradation and climate-related disasters. The chapter highlights South Africa's vulnerability to climate change, compounded by its socio-economic inequalities and high greenhouse gas emissions.

In Affonso's chapter, *The Impact of Global Boiling on Caribbean Islands: Practicality of Achieving Sustainable Development Goals*. The United Nations Sustainable Development Goals (SDGs) are seen as universal standards of global development that each country should aim for. The small island developing states (SIDS) of the Caribbean find it impossible to accomplish due to the challenges which are faced by the SIDS of the Caribbean including resource limitations, mobility obstacles, and sensitive economic relationships between the environment and livelihoods.

The implications of these challenges emphasize the need for context-specific approaches to sustainable development. SIDS require tailored frameworks that address their distinct environmental, economic, and social realities rather than relying on a one-size-fits-all model like the SDGs. This includes redefining sustainability in a way that balances economic growth with environmental conservation and prioritizing adaptive strategies that align with local capacities. Recommendations include addressing disparities, international support to SIDS through financial aid, technological assistance, and capacity-building initiatives.

Arneaud's chapter, *The psychosocial impact of climate change: A recent overview*, explores the psychosocial impacts of climate change, emphasizing its effects on human well-being and the increasing need for psychological and social interventions. Climate change poses immediate threats to physical health, socio-economic stability, and mental health. Rising global temperatures, natural disasters, and environmental degradation contribute to conditions such as post-traumatic stress disorder (PTSD), major depression, ecoanxiety, and heightened rates of self-harm and suicide. The chapter highlights the direct and indirect impacts of climate change on mental health, including acute stress from disasters and vicarious effects like ecoanxiety caused by awareness of environmental degradation. Children, Indigenous populations, and those living in climate-sensitive regions face disproportionate risks due to their reliance on natural resources and limited access to mental health resources.

This chapter advocates for a range of interventions such as therapy-based approaches that target acute and vicarious impacts, using cognitive and emotional strategies to build resilience. Community-based interventions emphasize social support and collective action to mitigate distress and foster connection. Climate literacy programs aim to educate and empower individuals to engage in proactive climate solutions, making information personally relevant to enhance agency and preparedness.

The implications for this section highlight the multilayered challenges posed by climate change and the urgent need for tailored interventions and policies. Both Nel and Khosa's and Affonso's chapters emphasize that universal frameworks, such as the Sustainable Development Goals (SDGs), are insufficient for addressing the unique vulnerabilities of regions like South Africa and the Caribbean SIDS. These areas require context-specific approaches that prioritize adaptive strategies, equitable resource distribution, and the integration of local capacities to address the compounded impacts of climate change.

The intersection of social work and climate justice, as discussed by Nel and Khosa, underscores the critical role social workers play in advocating for policies and mobilizing communities to address systemic inequities exacerbated by climate change. Similarly, Arneaud's chapter highlights the escalating mental health crisis triggered by climate-related stressors. This calls for therapy-based, community-centered, and educational interventions to integrate mental health strategies into climate policies. Together, these chapters stress the need for holistic, interdisciplinary solutions that combine environmental, social, and psychological strategies.

The second section delves deeply into the complex relationship between climate change and migration, examining how environmental disruptions drive displacement and reshape the lives of affected populations.

In Huggins' chapter, *The impact of climate change among displaced migrants in small island developing states: The case of Trinidad and Tobago*, addresses the multifaceted challenges faced by Small Island Developing States (SIDS), such as Trinidad and Tobago, in the context of climate change. The chapter highlights the significant social and environmental vulnerabilities of SIDS including poverty, crime, unemployment, and migration and how they intersect with climate-related challenges, creating compounded risks for these regions. Migrants and displaced individuals, particularly climate refugees, face heightened vulnerabilities due to limited resources, xenophobia, and inadequate access to education and healthcare in host countries like Trinidad and Tobago.

Trinidad and Tobago's government and international partners have made strides in addressing these issues, including implementing climate change policies and providing social welfare support to vulnerable populations. However, challenges persist due to resource constraints, colonial governance legacies, and the lack of a robust social welfare system. The chapter calls for a holistic approach to climate change and social welfare, emphasizing the need for equitable policies, enhanced adaptive capacity, and sustainable development strategies tailored to the unique needs of SIDS.

The Monico's chapter, *Mitigation and Adaptation Innovations Addressing Climate Change and Forced Migration: Case Study Research in Five World Regions. This chapter explores* climate change as an environmental disaster and a human emergency. Millions have become internally displaced or forced to emigrate from their countries of origin. Vulnerable populations are most impacted as they face structural and systemic oppression and possess limited resources to cope with climate disasters. This chapter summarizes findings on climate change impact on human livelihoods in Central and South America, Horn of Africa, South Asia, the Mediterranean Basin, and South-eastern Europe. Case study research suggests that climate change is a key. The chapter underscores the need for innovative mitigation and adaptation strategies that prioritize the well-being of climate-affected populations.

In the second section these chapters underscore the urgent need for targeted, inclusive approaches to address the intersecting challenges of climate change, migration, and social vulnerability in countries located in the global south. Huggins' chapter highlights the compounded risks faced by SIDS like Trinidad and Tobago, where social vulnerabilities such as poverty, crime, and unemployment intersect with climate-related challenges. Displaced migrants, particularly climate refugees, face heightened risks, including xenophobia, inadequate access to education and healthcare, and limited resources.

Monico's chapter further emphasizes the global dimension of climate-induced displacement, framing it as both an environmental disaster and a human emergency. Case studies from regions such as South America, the Horn of Africa, and South Asia highlight the role of climate change in disrupting livelihoods, driving migration, and undermining cultural continuity. Both chapters call for innovative mitigation and adaptation strategies that address the needs of displaced and climate-affected populations. These insights stress the importance of integrating climate change, migration, and social justice into global and local frameworks to create sustainable and inclusive responses to these pressing challenges.

The third section focuses on proactive interventions and innovative solutions designed to empower communities. It highlights strategies that foster resilience, sustainability, and adaptation, offering a roadmap for mitigating climate-related challenges.

In Nyeverwai Rumutsa's chapter, *Food security through entomophagy: The benefits and challenges of harvesting Encosternum delegorguei in climate-change-affected rural Nerumedzo community, Bikita, Zimbabwe.* This chapter explores alternative sources of food and income in communities that once depended on agriculture. Hararwa is harvested both for household consumption and as a source of income when sold. The findings suggest that, despite the challenges associated with harvesting hararwa, it contributes to improving household food security and boosting family incomes. The harvesting of *Encosternum delegorguei* (hararwa) is a sustainable solution to address climate change challenges due to its low environmental impact and adaptability. By offering an alternative income source less affected by agricultural disruptions, hararwa contributes to community resilience and aligns with the principles of sustainable food systems.

Shayamano, Masinga, and Dube's chapter, *Climate Change-Induced Disasters in Manicaland Province in Zimbabwe: The Need for Continuous Professional Development*, explores the social work profession in addressing climate change-related disasters, focusing on the Manicaland Province. The Manicaland Province have been experiencing an increased frequency and severity of meteorological disasters, such as cyclones, floods, and droughts, which have significantly impacted communities with destroyed infrastructure, displaced populations and a lack of resources. Although, social work practitioners are primarily engaged in disaster response and recovery, providing immediate relief, counseling, and support to affected communities, they lack training and exposure to disaster-focused curricula in the social work curriculum.

The chapter emphasizes the need for integrating climate change and disaster management into social work education and creating a Continuous Professional Development (CPD) program. These types of programs are not available in this region, but it is critical for practitioners to learn critical thinking skills such as disaster planning, crisis intervention, and resource mobilization. Ultimately, the chapter underscores the importance of empowering social workers to support communities in building resilience and adapting to the challenges posed by climate change-induced disasters.

Moore, Hinkson, and Clarke's chapter, *Climate Change and its Effect on the Mental Health of Farmers in Trinidad & Tobago*. This chapter examines the mental health of farmers in the context of one of the Caribbean islands vulnerable to floods and other climate change-related effects. While the impact of climate change on agriculture and livelihoods has been extensively studied, the link between farmers' mental health and climate change has received little attention. The chapter emphasizes the importance of prioritizing interventions that address farmers' mental health needs and improving access to resources that help them cope with the negative effects of climate change, such as crop loss, drought, and floods. The findings highlight the need for an integrated approach to climate change mitigation and adaptation that includes mental health as a critical component. Addressing farmers' mental health not only improves individual well-being but also enhances resilience, enabling farmers to better cope with the negative impacts of climate change, such as crop loss, drought, and floods.

The Zibane, Phoswa, and Zondi, *Centering the response to the KwaZulu-Natal floods in the African-centred Psychosocial Interventions* chapter interrogates the 2022 KwaZulu-Natal floods in South Africa and their devastating impacts, including significant loss of life, displacement, and trauma among affected communities. This chapter highlights the limitations of conventional Eurocentric trauma responses and emphasizes the need for culturally relevant interventions. Grounded in the African-centered worldview, the Ma'at Institute implemented African-centered Psychosocial Interventions (APIs) to aid survivors. The chapter outlines key principles for engaging African Indigenous Communities (AICs), including building trust through culturally appropriate entry strategies, using indigenous languages, and fostering rapport through shared cultural practices. These interventions focused on healing, resilience building, and restoring dignity through culturally specific tools and techniques. The chapter underscores the importance of integrating indigenous knowledge and cultural practices into disaster recovery. APIs employed methods like drumming, music, storytelling, and communal healing activities, which resonated deeply with survivors' cultural identities. These tools addressed not only physical and emotional well-being but also the social and spiritual dimensions of recovery, fostering a holistic approach to healing. Collaborative autoethnography was used to reflect on the interventions and document the significance of honoring local knowledge and traditions.

This chapter advocates for centering disaster responses within the cultural and historical contexts of affected communities. By prioritizing African ethics, indigenous knowledge systems, and community-driven approaches, the APIs demonstrated the transformative power of culturally relevant psychosocial interventions in addressing trauma and fostering resilience in disaster-affected communities.

Shoyne's chapter, *The Climate-Resilient Elderly of Barbados,* investigates how older adults' resilience in two flood-prone communities in Belleplaine and Holetown Barbados. Using hermeneutic phenomenology and resilience theory, this study interviewed 8 participants on their lived experiences and coping strategies. Three key themes emerge: trust in God, local knowledge of community geography, and reliance on social support networks and government interventions. Elderly participants demonstrated strong faith, viewing spirituality as a source of strength and reassurance during flood events. Their deep knowledge of local watercourses and flood-prone areas underscores the importance of incorporating indigenous knowledge into disaster management strategies. Social support from family, friends, and neighbors plays a crucial role, often supplementing government recovery efforts, which were sometimes perceived as insufficient.

The study underscores the importance of recognizing the elderly as active agents in resilience, not merely as a vulnerable group. Their lived experiences and local knowledge are invaluable in shaping disaster preparedness and recovery plans. The findings call for more inclusive approaches to disaster management, highlighting the interplay of personal, communal, and institutional factors in fostering resilience. This study advocates for further research into elderly resilience to climate related events across different communities to inform policies and practices that support this growing demographic.

In Bain and Low's chapter, *The power of small-scale interventions in progressing climate adaptation and resilience in the Caribbean: A case study of Trinidad and Tobago.* This chapter examines climate resilience in Trinidad and Tobago, emphasizing the role of micro, small, and medium enterprises (MSMEs) in addressing climate change. MSMEs, which constitute 80% of registered businesses and contribute significantly to GDP, are highlighted as underutilized yet critical players in climate action. These enterprises leverage local knowledge and grassroots involvement, providing context-specific, sustainable solutions. Unlike top-down government or corporate strategies, MSMEs emphasize collective action and community-centered approaches. However, their potential is limited by misconceptions about their willingness to invest in sustainability and the lack of supportive government policies. This chapter calls for a shift towards inclusive, localized climate action, integrating MSMEs into national policies and supporting their capacity to address climate challenges.

Climate Models using AI: A case for the OECS was written **by** Letetia M. Addison. This chapter discusses disaster preparedness due to climate change as a critical component in Caribbean islands. In particular, countries in the Global South such as those in the Organization of Eastern Caribbean States (OECS) are prone to natural disasters annually due to their geographic location. Increased climate resilience, through proper planning mechanisms is a key factor in the mitigation of damages and loss of life due to these disasters. In most cases, the necessary data may be available but inaccessible to vulnerable communities. AI techniques are essential to improve disaster preparedness and response mechanisms via proper coordination and activation of information directed to emergency services. In this work, we review current AI modeling techniques which have been used for this purpose and present a case for the OECS regions using country-level climate variables to predict the risk of a natural disaster. These can be built into scalable models to provide key stakeholders with critical insights and enhanced tools using data-driven approaches for efficient disaster management response in these regions.

The chapter emphasizes the transformative role of AI in improving disaster preparedness and response in the Organization of Eastern Caribbean States (OECS) and other Global South nations, which are highly vulnerable to climate-related disasters. By integrating AI-driven climate models, these regions can significantly enhance resilience, mitigate damage, and reduce loss of life. AI offers the capability to process and disseminate critical climate data, much of which is currently inaccessible to vulnerable communities. This ensures that actionable insights reach those who need them most, empowering communities to better prepare for disasters.

This section highlights a wide range of implications for addressing climate change, emphasizing the importance of innovative, context-specific approaches that integrate environmental, social, and technological solutions to foster resilience and adaptation. Several chapters underscore the value of integrating indigenous knowledge and community-based approaches into climate resilience strategies. Rumutsa's chapter highlights the importance of sustainable practices, like harvesting *Encosternum delegorguei*, while Shoyne's chapter emphasizes the critical role of elderly populations' local knowledge in disaster preparedness. Similarly, Zibane et al.'s chapter showcase the transformative power of African-centered psychosocial interventions that incorporate cultural practices and community-driven healing. Some of the chapters called for more social work presence in the climate change response. Both Rumutsa and Shayamano et al.'s chapters advocate for embedding climate change and disaster management into social work practice and education. Moore et al.'s chapter highlights the often-overlooked mental health impacts of climate change, particularly among farmers. It is important to integrate mental health interventions into broader climate adaptation strategies, recognizing the interplay between individual well-being and community resilience. Bain and Low's chapter emphasize the untapped potential of micro, small, and medium enterprises (MSMEs) in driving localized climate action in Trinidad and Tobago. Similarly, Addison underscores the transformative role of AI-driven models in enhancing disaster preparedness for vulnerable regions like the OECS, stressing the need for accessible, scalable technologies.

In conclusion these chapters collectively advocate for inclusive, localized, and interdisciplinary approaches that integrate traditional knowledge, community-driven initiatives, and modern technologies to effectively address the complex challenges posed by climate change.

Index

D. D. Joseph et al. (eds.), *Climate Change*,
https://doi.org/10.1007/978-3-031-83037-2

The manufacturer's authorised representative in the EU is Springer Nature Customer Service Centre GmbH, Europaplatz 3, 69115 Heidelberg, Germany. If you have any concerns regarding our products, please contact ProductSafety@springernature.com

Printed and bound by CPI Group (UK) Ltd, Croydon, CR0 4YY
30/09/2026
02225565-0013